T0140840

Design of Distributed and Robust Optimization Algorithms

A Systems Theoretic Approach

Von der Fakultät Konstruktions-, Produktions- und Fahrzeugtechnik
der Universität Stuttgart zur Erlangung der Würde eines
Doktor-Ingenieurs (Dr.-Ing.) genehmigte Abhandlung

Vorgelegt von

Simon Michalowsky

aus Stuttgart

Hauptberichter: Prof. Dr.-Ing. Christian Ebenbauer
Mitberichter: Prof. Mohamed-Ali Belabbas, Ph. D.
Prof. Dr. rer. nat. Carsten W. Scherer

Tag der mündlichen Prüfung: 28. August 2019

Institut für Systemtheorie und Regelungstechnik
der Universität Stuttgart

2019

Bibliografische Information der Deutschen Nationalbibliothek

Die Deutsche Nationalbibliothek verzeichnet diese Publikation in der Deutschen Nationalbibliografie; detaillierte bibliografische Daten sind im Internet über http://dnb.d-nb.de abrufbar.

D 93

© Copyright Logos Verlag Berlin GmbH 2020
Alle Rechte vorbehalten.

ISBN 978-3-8325-5090-5

Logos Verlag Berlin GmbH
Comeniushof, Gubener Str. 47,
D-10243 Berlin
Germany

Tel.: +49 (0)30 / 42 85 10 90
Fax: +49 (0)30 / 42 85 10 92
http://www.logos-verlag.de

Acknowledgements

This thesis is the outcome of my time as a research and teaching assistant at the Institute for Systems Theory and Automatic Control (IST) at the University of Stuttgart. I want to express my gratitude to all people who were part of this very instructive and pleasant period.

I primarily would like to thank my advisor Prof. Christian Ebenbauer for giving me the chance to work with him, his guidance and his valuable feedback. I am particularly grateful for the numerous discussions, his advice and the way he encouraged me to pursue challenging research projects. My gratitude extends to all other members of the examination committee, namely Prof. Mohamed-Ali Belabbas, Prof. Carsten W. Scherer and Prof. David Remy. I additionally would like to thank Prof. Carsten W. Scherer for arousing my interest in robust control and the very valuable discussions. Many thanks also go to Prof. Bahman Gharesifard for inviting me to Queen's University and the productive and pleasant collaboration.

I further want to express my thanks to Prof. Frank Allgöwer for establishing such a fruitful academic environment at the IST and the possiblities and freedom he offers to his employees. Special thanks go to all my former colleagues and fellow PhD students of which many became friends. They contributed a lot to making the time at the IST so enjoyable and I will always keep it in best memory.

A heartfelt thank goes to my family for always supporting me and to my partner Amelie for her support, her understanding and her love.

Stuttgart, February 2020
Simon Michalowsky

Table of Contents

Abstract

In this thesis, we consider the analysis and design of first-order optimization algorithms employing systems and control theory. We recast algorithm design problems as controller synthesis problems; techniques from control theory then enable us to systematically construct tailored optimization algorithms adapted to various situations. In particular, we consider two specific classes of optimization algorithms: (i) continuous-time distributed optimization algorithms for constrained convex optimization, and (ii) robust discrete-time optimization algorithms for unconstrained convex optimization.

Concerning (i), we consider a group of agents sharing information over a communication network described by a directed time-invariant graph aiming to cooperatively solve a convex optimization problem with shared equality and inequality constraints. Utilizing geometric control theory in a novel and innovative fashion, in particular Lie bracket averaging techniques, we directly address the core challenge of distributed problems, namely limited local information. Employing saddle-point dynamics, we derive a novel methodology that enables the design of distributed continuous-time optimization algorithms solving a class of optimization problems under minimal assumptions on the graph topology as well as on the structure of the optimization problem. Generalizing this approach, we further establish a systematic way of deriving continuous-time distributed algorithms from non-distributed ones.

Concerning (ii), we consider the problem of analyzing and designing gradient-based discrete-time optimization algorithms for a class of unconstrained optimization problems having strongly convex objective function with Lipschitz continuous gradient. By formulating the problem as a robustness analysis problem and employing a suitable adaptation of the theory of integral quadratic constraints (IQCs), we establish a framework that allows analyzing convergence rates and robustness properties of existing algorithms and enables the design of novel robust optimization algorithms with specified guarantees. Taking advantage of the embedding into integral quadratic constraint theory, we further extend the framework to design algorithms that are capable of exploiting additional structure in the objective function.

Deutsche Kurzfassung

In der vorliegenden Arbeit werden Analyse- und Entwurfsmethoden für gradientenbasierte Optimierungsalgorithmen entwickelt, indem mit Hilfe von systemtheoretischen Ansätzen das Entwurfsproblem in geeigneter Weise als ein Reglerentwurfsproblem umformuliert wird. Dieser Ansatz ermöglicht es, auf systematische Art und Weise maßgeschneiderte Optimierungsalgorithmen zu entwerfen. Dabei werden in dieser Arbeit insbesondere zwei spezielle Klassen von Optimierungsalgorithmen betrachtet: (i) zeitkontinuierliche verteilte Optimierungsalgorithmen zur Lösung konvexer Optimierungsprobleme mit Nebenbedingungen und (ii) robuste zeitdiskrete Optimierungsalgorithmen zur Lösung konvexer Optimierungsprobleme ohne Nebenbedingungen.

In Fall (i) verfolgt eine Gruppe von Recheneinheiten, auch Agenten genannt, das Ziel, kooperativ ein Optimierungsproblem mit gemeinsamen Gleichungs- und Ungleichungsnebenbedingungen zu lösen, wobei jeder Agent nur Zugriff auf eine begrenzte Menge lokal verfügbarer Information hat. Dazu tauschen die Agenten untereinander diese lokalen Informationen über ein Kommunikationsnetz aus, das abstrakt durch einen gerichteten, zeitinvarianten Graph repräsentiert werden kann. Der vorgestellte Ansatz basiert auf einer neuartigen und innovativen Verwendung von Methoden aus der geometrischen Regelung, im Speziellen Lie Klammer Approximationen. Dieser Ansatz erlaubt es, die Hauptschwierigkeit verteilter Probleme, nämlich die nur lokal verfügbare Information, direkt und systematisch anzugehen. Eine Kombination der Methodik mit Sattelpunktdynamiken ermöglicht es dann, verteilte zeitkontinuierliche Optimierungsalgorithmen unter geringen Voraussetzungen an die Struktur des Graphen und des Optimierungsproblems zu entwerfen. Durch weitere Verallgemeinerung dieser Methodik wird zudem ein systematischer Ansatz zum Entwurf verteilter aus nicht verteilten Algorithmen vorgestellt.

In (ii) wird die Analyse und der Entwurf gradientenbasierter zeitdiskreter Optimierungsalgorithmen für eine Klasse von unbeschränkten Optimierungsproblemen betrachtet, deren Kostenfunktion stark konvex ist und einen Lipschitz-stetigen Gradienten besitzt. Durch Umformulierung des Entwurfproblems als ein Problem der robusten Regelung und eine geeignete Modifikation der sogenannten IQC-Theorie (integral quadratic constraints) wird eine Methodik hergeleitet, die es erlaubt, sowohl die Konvergenzraten und Robustheitseigenschaften von bekannten Algorithmen zu analysieren als auch neue Optimierungsalgorithmen zu entwerfen, die vorgegebene Konvergenzraten- und Robustheitsgarantien erfüllen. Die Einbettung in die IQC-Theorie ermöglicht es dabei auch, Algorithmen zu entwerfen, die mögliche strukturelle Eigenschaften der Kostenfunktion explizit ausnutzen können.

1 Introduction

1.1 Motivation and Background

Optimization plays an important role in many fields of applications and is the backbone of a multitude of modern technologies such as real-time control, machine learning or data analytics. Having reliable, fast and flexible optimization algorithms available hence is of key importance. Over the last decades, a variety of optimization algorithms applicable in different situations have been developed and proven themselves in real-world applications. However, the technological progress often requires the adaptation to novel challenges that existing algorithms cannot handle appropriately, hence necessitating the development of tailored algorithms. Though, it is fair to say that the design of new algorithms, but also their analysis, still is more art than science based on experience, expert knowledge and good ideas. A systematic framework to analyze and modify existing or design novel algorithms adapted to different situations is not available yet. On the other hand, many optimization algorithms are, in essence, dynamical systems having an equilibrium characterized by the solutions of a class of optimization problems. Systems and control theory provides a quite mature set of tools for analyzing the convergence and stability properties of equilibria as well as for designing controllers that stabilize a given equilibrium. While the apparent relation between these two areas of research is not a new discovery, its full potential has not been exploited yet.

In the present thesis, we aim to contribute to linking the areas of optimization and systems and control theory and show that the latter provides novel ways to analyze and design optimization algorithms adapted to various problem setups. The main theme is to recast optimization algorithm design problems as particular controller synthesis problems and utilize methods from systems and control theory to enable a systematic design of optimization algorithms, see also Figure 1.1. In this thesis, we consider two particular classes of algorithm design problems, namely distributed optimization and robust optimization in the presence of noise; we are convinced that systems and control theory has the potential to provide a powerful approach to optimization algorithm design in general.

This view is supported by a growing amount of publications and applications following this systems theoretic approach to optimization. Recent advances in machine learning, data science or real-time decision-making as well as optimization-based control techniques such as model predictive control rely to a large extent on efficient optimization algorithms.

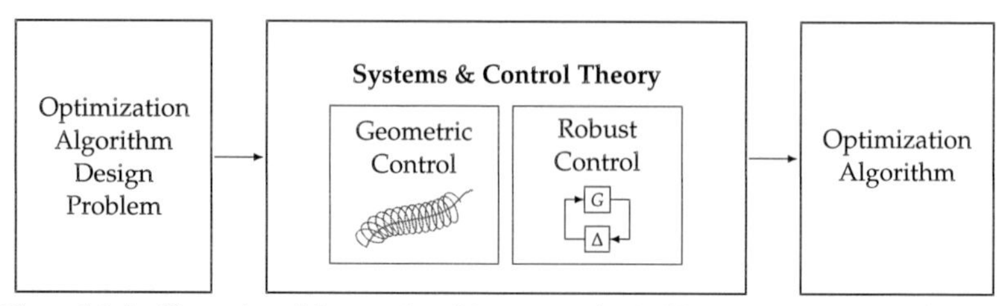

Figure 1.1. An illustration of the premise of the present thesis: We recast optimization algorithm design problems as controller synthesis problems and utilize and extend methods from systems and control theory, in particular geometric control and robust control, to facilitate a systematic design of tailored optimization algorithms.

This development has pushed the need for optimization algorithm design tools, which in turn motivated researchers to more thoroughly investigate and exploit the link between optimization algorithms and systems theory in the last years. Historically, this link can be traced back to the early days of Extremum Seeking Control (Draper & Li, 1951; Leblanc, 1922). While several other results from the last century also can be said to follow similar system theoretic ideas for optimization algorithm design, it was in Brockett (1988, 1991) where the author explicitly utilized geometric control theory to design dynamical systems that are able to solve tasks such as sorting lists or solving linear programs, providing an alternative to classical algorithms for such type of problems. This approach was further pursued by a group of researchers and other tasks, e.g., singular value decomposition, were addressed from a systems theoretic perspective as well (Helmke & Moore, 1994). Still, even more than ten years later, it is stated in Bhaya and Kaszkurewicz (2006) that although "some simple ideas from control theory can be used to systematize a class of approaches to algorithm analysis and design [...] control and system theory ideas have been underexploited in this context". In the last decade, several authors picked up on that (Dürr & Ebenbauer, 2012; Hauswirth, Bolognani, Hug, & Dörfler, 2019; Michalowsky & Ebenbauer, 2014; Wang & Elia, 2010, 2011, just to name a few), mainly from a continuous-time perspective that is probably more common in control theory, which also lead to a regained interest in continuous-time optimization algorithms (Gharesifard & Cortés, 2014; Niederländer & Cortés, 2015; Su, Boyd, & Candes, 2014). Recent advances (Wibisono, Wilson, & Jordan, 2016) further promote this continuous-time perspective enabling a deeper understanding also of discrete-time algorithms. Another approach that turned out to be a very fruitful example for utilizing control theoretic methods in optimization is the interpretation of gradient-based optimization algorithms in a robust control setting: with optimization algorithms typically having to be applicable to a class of objective functions, the idea of interpreting the gradient of the objective function as an uncertainty seems natural. This idea was first followed in Michalowsky and Ebenbauer (2014) in a continuous-time setting and in Lessard, Recht, and Packard (2016) for discrete-time optimization. In the latter work, the authors utilize integral quadratic constraint theory, which is well-established in robust control, to analyze a class of optimization algorithms, thereby unifying the analysis of several popular gradient-based algorithms. In the last two years, this approach has been followed in a number of publications (e.g., Aybat, Fallah, Gürbüzbalaban, and Ozdaglar (2019); Cyrus, Hu, Van Scoy, and Lessard (2018); Fazlyab, Ribeiro, Morari, and Preciado (2018); Michalowsky, Scherer, and Ebenbauer (2020); Safavi, Joshi, França, and Bento (2018); Van Scoy, Freeman, and Lynch (2018)).

1.2 Problem Formulation

In this thesis, we aim to provide a systems theoretic approach to two particular aspects of optimization: (i) distributed optimization over directed graphs and (ii) fast and robust optimization in the presence of noise. Following the premise illustrated in Figure 1.1, we address these two aspects by a proper reformulation of the algorithm design problem as a controller synthesis problem. We formalize the specific problems in the remainder and briefly sketch how we approach them.

Throughout the thesis, we consider (constrained) convex optimization problems

$$\begin{aligned} \underset{z\in\mathbb{R}^p}{\text{minimize}} \quad & H(z) \\ \text{s.t.} \quad & a(z) = 0 \\ & c(z) \leq 0, \end{aligned} \tag{1.1}$$

where $H : \mathbb{R}^p \to \mathbb{R}$, $a : \mathbb{R}^p \to \mathbb{R}^{n_{\text{eq}}}$, $c : \mathbb{R}^p \to \mathbb{R}^{n_{\text{ineq}}}$, $n_{\text{eq}}, n_{\text{ineq}} \in \mathbb{N}_{>0}$, $H, a, c \in \mathcal{C}^2$. Our goal is then to design optimization algorithms, i.e., dynamic systems, both in continuous- and discrete-time, that converge to a minimizer of (1.1), assuming that such a minimizer exists. More precisely, we consider deterministic optimization algorithms described by continuous- or discrete-time dynamic systems

$$x^+(t) = f\big(t, x(t)\big) \tag{1.2a}$$
$$z(t) = h\big(x(t)\big), \tag{1.2b}$$

where $x(t) \in \mathbb{R}^N$ for some $N \geq p$, $z(t) \in \mathbb{R}^p$, and f, h are functions to be designed. In a continuous-time setting, t is a non-negative real number representing the time and $x^+(t) = \dot{x}(t) = \frac{\mathrm{d}x}{\mathrm{d}t}(t)$ is the usual time derivative; in a discrete-time setting, t takes integer values representing time steps and $x^+(t) = x(t+1)$. We concentrate on first-order optimization algorithms, i.e., f may only depend on $H, a, c, \nabla H, \nabla a, \nabla c$ but no higher-order derivatives. In simple words, our overall goal is then formulated as follows:

General Optimization Algorithm Design Problem. "Given a class of optimization problems of the form (1.1) and some design specifications, design optimization algorithms (1.2) such that, as t tends to infinity, $z(t)$ converges to a minimizer of (1.1) for any instance of (1.1) and the specifications are met."

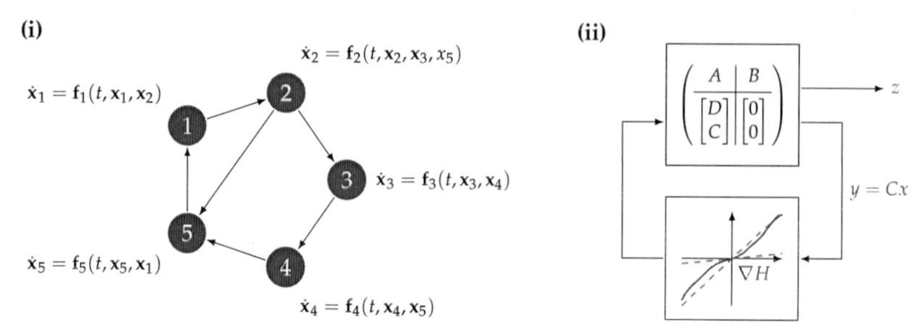

Figure 1.2. Two exemplary problems of type (i) and (ii). **Left (i):** A communication network of $n = 5$ agents with agent states $\mathbf{x}_i$, $i = 1, \ldots, 5$. The arrows indicate the directions of information access; for example, in addition to its own state, agent 2 has access to the states $\mathbf{x}_3, \mathbf{x}_5$ of agent 3 and 5 but not to the states of the other agents in the network. This is reflected in the agent dynamics depicted next to the agents and illustrates the main challenge in designing distributed algorithms, namely limited locally available information. **Right (ii):** A block diagram representation of the class of algorithms (1.5) we consider in (ii). We reformulate the algorithm design problem as a robust controller synthesis problem interpreting the gradient of the objective function ∇H as an uncertainty. The challenge is then to design matrices A, B, C, D such that, as t tends to infinity, $z(t)$ converges to a minimizer of the optimization problem (1.1) for all H in some class.

We address the latter problem by designing the functions f, h in such a way that (1.2) has a (globally) asymptotically stable equilibrium at a point $x^\star$ that has the property that $h(x^\star) = z^\star$, where $z^\star$ is a minimizer of (1.1). In this manner, in its core the problem boils down to a stabilization problem with the important distinction that the point to be stabilized is not known a-priori but determined by (1.1). As to the additional design specifications, we are particularly interested in addressing two specific aspects of this general problem: (i) distributed optimization over directed graphs and (ii) fast and robust optimization in the presence of noise. Those two aspects are highly relevant in modern applications; we make this more precise in the following and specify the general problem formulation for (i) and (ii).

(i) Distributed Optimization Algorithms. In distributed optimization, a group of computation units, often called agents, cooperatively tries to solve an optimization problem. The idea of distributed algorithms is to have each agent solve a smaller subproblem using a limited amount of local information only and, by sharing information amongst the agents over a communication network, ensure that the original problem is solved. Distributed algorithms have the advantage that they are usually less error-prone, might require less communication and can respect possible privacy issues.

In this part, we are aiming for designing distributed continuous-time optimization algorithms (1.2). More precisely, we consider a group of n agents where each agent's state evolves according to its individual agent dynamics described by a differential equation and the complete distributed algorithm (1.2) is given by the collection of all agent dynamics. There is no common understanding of a distributed algorithm; in this thesis, simply put, we call an algorithm distributed if each agent only uses its own state as well as the states of the agents it has access to, where information access is encoded by a graph representation of the communication network. An exemplary illustration of the situation is depicted in Figure 1.2, (i). As visible from this example, in essence, this understanding of a distributed algorithm then amounts to design f in (1.2) in such a way that it respects certain information constraints induced by the communication network. More precisely, we consider a subclass of the general continuous-time algorithm dynamics (1.2) given by

$$\begin{bmatrix} \dot{x}_1(t) \\ \dot{x}_2(t) \\ \vdots \\ \dot{x}_n(t) \end{bmatrix} = \begin{bmatrix} f_1(t, [x_k(t)]_{k \in \mathcal{N}_\mathcal{G}(1)}) \\ f_2(t, [x_k(t)]_{k \in \mathcal{N}_\mathcal{G}(2)}) \\ \vdots \\ f_n(t, [x_k(t)]_{k \in \mathcal{N}_\mathcal{G}(n)}) \end{bmatrix} \tag{1.3a}$$

$$z(t) = h(x(t)), \tag{1.3b}$$

where $x_i(t) \in \mathbb{R}^{N_i}$ is the state of the ith agent and $[x_k]_{k \in \mathcal{N}_\mathcal{G}(i)}$ is the collection of states of agents the ith agent has access to via the communication network, i.e., in a graph theoretic language, the out-neighboring states of agent i as well as its own state. Our goal is then to design the functions f_i in (1.3) such that $z(t)$ converges to a minimizer of (1.1). The main challenge in designing distributed algorithms (1.3) compared to general algorithms (1.2) is the limited information available to each agent rendering the stabilization problem a sparse stabilization problem. In other words, from a controller design perspective, there is a limited set of admissible control directions determined by the communication network. Such situations have been investigated a lot in nonlinear control theory, in particular in geometric control, e.g., in controllability analysis or the control of nonholonomic mechanical systems. Our approach relies on employing these ideas in a novel way to design distributed optimization algorithms.

(ii) Robust Optimization Algorithms. In many applications it is of key importance to have optimization algorithms that provide an accurate solution in guaranteed time. However, the convergence time of optimization algorithms is often hard to analyze and further, existing algorithms may not fulfill the imposed convergence rate requirements. Besides, while many algorithms perform well in an idealized setting, they are sensitive towards various disturbances, resulting in slow convergence. Such kind of disturbances arise, for example, in a data-based setting where the optimization problem specifiers H, a, c are generated from data.

In the second problem investigated in this thesis we are concerned with the design of discrete-time optimization algorithms applicable to a class of unconstrained optimization problems that (a) provide specified convergence rate guarantees, (b) are insensible towards noise in the optimization problem data (i.e., H in (1.1)) and (c) are capable of exploiting additional structural properties of the objective function. The class of algorithms we propose in this thesis is motivated by a generalization of existing algorithms. We explain the idea by means of the Heavy Ball Method (Polyak, 1987) that, for scalar optimization ($p = 1$), can be represented in the form (1.2) as

$$\begin{bmatrix} x_1(t+1) \\ x_2(t+1) \end{bmatrix} = \begin{bmatrix} 1+\nu_2 & -\nu_2 \\ 0 & 1 \end{bmatrix} \begin{bmatrix} x_1(t) \\ x_2(t) \end{bmatrix} - \begin{bmatrix} \nu_1 \\ 0 \end{bmatrix} \nabla H(x_1(t)) \tag{1.4a}$$

$$z(t) = x_1(t), \tag{1.4b}$$

where ν_1, ν_2 are real parameters. A suitable choice of parameters for a class of objective functions H has been derived in (Polyak, 1987) where also the convergence of the algorithm (1.4) towards a minimizer $z^\star$ of H has been analyzed. The following question arises: How do these parameters need to be adapted – and how does this affect the convergence rate – if the class of objective functions changes or is refined or if the gradient of the objective function is affected by noise? Going further, how do we need to change (1.4) structurally in order to adapt to such novel situations? In this part of the thesis, we provide answers to such questions by generalizing (1.4) to a particular subclass of optimization algorithms (1.2) of the form

$$x(t+1) = Ax(t) + B\nabla H(Cx(t)) \tag{1.5a}$$

$$z(t) = Dx(t), \tag{1.5b}$$

where $x(t) = \begin{bmatrix} x_1(t)^\top & \dots & x_n(t)^\top \end{bmatrix}^\top \in \mathbb{R}^{np}$, $x_i(t) \in \mathbb{R}^p$, $i \in \{1, \dots, n\}$, $z(t) \in \mathbb{R}^p$. The design goal then amounts to determine matrices $A \in \mathbb{R}^{np \times np}$, $B \in \mathbb{R}^{np \times p}$, $C \in \mathbb{R}^{p \times np}$, $D \in \mathbb{R}^{p \times np}$, independent of H, such that, for any H in a given class, the dynamics (1.5a) have an asymptotically stable equilibrium at $x^\star$ with the property that $Dx^\star = z^\star$, where $z^\star$ is the minimizer of H. The core idea is to interpret the unknown objective function – or more precisely its gradient – as an uncertainty (see Figure 1.2, (ii)); we then need to render $x^\star$ asymptotically stable for (1.5) for any realization of the uncertainty. In other words, the problem is recast as a robust controller synthesis problem.

1.3 Contributions and Outline

In this thesis, we provide systematic procedures to analyze and design two classes of optimization algorithms. In particular, following the premise of a systems theoretic approach, we embed both problems (i) and (ii) described in Section 1.2 in a systems and control theoretic setup. For each problem (i), (ii), we then provide a framework that applies to a large class of optimization problems, can be extended systematically and allows for an automation of the algorithm design process. The two problems also build the two main chapters of this thesis that we outline in the following.

- In Chapter 2, we address problem (i) and provide a systematic framework to derive continuous-time distributed optimization algorithms from non-distributed ones. The approach is based on Lie bracket averaging techniques. We propose a two-step procedure

where the first step consists of finding certain Lie bracket representations of parts of the algorithm that cannot be implemented in a distributed fashion (we call them non-admissible vector fields) and the second step is to determine distributed approximations thereof. We discuss the procedure by means of saddle-point dynamics and show how it can be applied to other non-distributed optimization algorithms.
More specifically, concerning the first step, we provide Lie bracket representations of a large class of non-admissible vector fields (Proposition 1, Lemma 3). Applying the results to saddle-point dynamics, we show that such Lie bracket representations can be obtained under mild assumptions on the optimization problem (Lemma 4). Concerning the second step of determining distributed approximations, we modify the construction procedure from Liu (1997a) and derive a simplified version thereof allowing an explicit representation of the distributed approximation in certain cases (Proposition 2). We finally combine both steps and thereby provide a methodology to derive distributed optimization algorithms from non-distributed ones (Theorem 2). The results of this chapter are based on the papers Michalowsky, Gharesifard, and Ebenbauer (2017a, 2018, 2020) and some parts of the text, in particular in Section 2.4, are identical.

- In Chapter 3, we address problem (ii). By formulating the problem as a robustness analysis problem and making use of a suitable adaptation of the theory of integral quadratic constraints, we establish a framework that allows to analyze convergence rates and robustness properties of existing algorithms and design novel optimization algorithms that are robust towards noise, fulfill specified guarantees and are capable of exploiting additional structure in the objective function.
Specifically, our main contributions are as follows: We propose a class of gradient-based algorithms that generalizes existing algorithms and derive necessary and sufficient conditions for these algorithms to be capable of solving a class of optimization problems (Theorem 3). Embedding the problem in the framework of robust control, we then derive convex analysis tools by means of linear matrix inequalities (LMIs), both in regard to convergence rates (Theorem 7) and robustness (Theorem 8). To this end, we provide a general procedure to obtain multipliers for exponential stability results from standard ones (Lemma 7) and utilize this to derive a class of multipliers generalizing those proposed in Boczar, Lessard, and Recht (2015); Freeman (2018); Lessard et al. (2016) (Theorem 6). We further provide convex synthesis conditions allowing the design of novel algorithms with specified robustness properties (Theorem 9) and show how to additionally exploit structural characteristics of the objective function (Lemma 10). The results presented in this chapter have been submitted to a large extent (Michalowsky, Scherer, & Ebenbauer, 2020) and are partly based on Michalowsky and Ebenbauer (2014, 2016).

- In Chapter 4, we give a summary of our results and discuss future research directions.

To streamline the presentation, we introduce the notation as well as the required technical background in a summarized form in Appendix A; an overview of the notation is also provided on page 147. All technical proofs of the mathematical statements are collected in Appendix B; Appendix C contains some additional material.

2 Design of Distributed Optimization Algorithms

Nowadays, nearly all devices we use in our daily life are equipped with microprocessors and connected to a network, be it cars, smartphones or fridges. This ubiquity of computational power and the growing interconnectedness opens up new possibilities but also novel challenges have to be faced. In particular, limitations in communication and questions of privacy lead to limited locally available information. This lack of information is a major difficulty to be addressed and distributed algorithms are designed as a remedy to such problems. With optimization being one of the key enablers of modern technology, the idea of solving optimization problems in a distributed fashion also got in the focus of interest in the last decades. Therein, a group of computation units (often also called *agents*) cooperatively tries to solve the problem. The idea of many distributed optimization algorithms is to have each agent solve a smaller subproblem and, by sharing information over a communication network, ensure that the original problem is solved.

Many of the existing approaches to distributed optimization problems heavily rely on the assumption that the underlying communication network is of undirected nature, meaning that when one agent shares information with a second agent, the second agent will share his information as well. However, many practical problems do not have this property, e.g., due to the directed nature of sensors in networks of physical agents or due to privacy reasons. Further, the available distributed algorithms typically require strong assumptions on the structure of the optimization problem, e.g., that the objective function is a sum of individual objective functions of each agent only or that the constraints are only imposed between agents that also share information with each other. This heavily limits the class of optimization problems these algorithms are able to solve or necessitates modifications of the existing communication network. Existing approaches trying to address and relax these limitations typically utilize specifically tailored modifications of a distributed algorithm, but no general approach is known in literature.

In this thesis, we aim to take a different approach and establish a framework that allows to systematically design continuous-time distributed optimization algorithms. From a more general perspective, this thesis also provides a systematic procedure for deriving distributed algorithms from non-distributed ones. Our methodology is applicable to a quite general class of convex optimization problems under rather mild assumptions on the com-

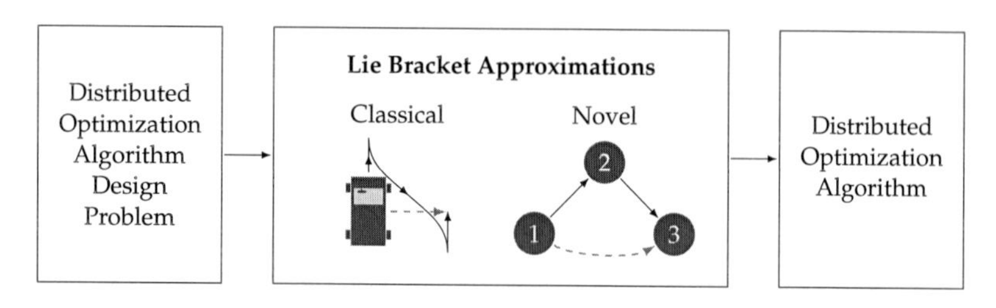

Figure 2.1. An illustration of our approach. We use Lie bracket approximations to systematically derive distributed algorithms from non-distributed ones. We employ Lie bracket approximation techniques which have proven useful to reveal and utilize hidden control directions (dashed) of nonlinear systems. Our approach further advances into that direction, where the hidden control direction is identified with a non-existing communication link (dashed).

munication structure. Following the main premise of this thesis, our approach builds upon well-established tools from systems and control theory. In particular, we use Lie bracket approximation techniques in a novel and innovative fashion. Lie bracket approximations have been extensively used in nonlinear control theory, e.g., in motion planning problems for nonholonomic mechanical systems (Z. Li & Canny, 2012, and references therein), where they enable steering a system into directions not directly accessible. We utilize Lie bracket approximations in a similar way in the sense that they enable an agent to use information somewhere available in the network but not directly accessible via a communication link (see Figure 2.1 for an illustration).

Background and Related Work. Over the last decades, distributed optimization and the closely related field of distributed control has been a very active area of research with high practical relevance, see, e.g., Boyd, Parikh, Chu, Peleato, and Eckstein (2011); Bullo, Cortés, and Martínez (2009); Zhao and Dörfler (2015) for applications. In the following, we give a brief overview of research in distributed optimization; due to the sheer amount of literature available in this area, we do not aim for a complete overview but refer the reader to the recent survey paper Nedić and Liu (2018). In distributed optimization, usually optimization problems (1.1) are considered, where, additionally, the objective function H is assumed to be a sum of individual objective functions associated to each of the n agents. More precisely, it is assumed that

$$H(z) = \sum_{i=1}^{n} H_i(z), \tag{2.1}$$

where $H_i : \mathbb{R}^p \to \mathbb{R}$ is the individual objective function associated to agent i. The goal is to solve the problem by dividing it into n smaller subproblems; each agent then solves a part of the original problem (1.1) and communicates its solution over a given communication network. Existing algorithms following such an approach can mainly be distinguished by means of (i) how the problem is divided into subproblems and (ii) the characteristics of the algorithms that solve these local subproblems. Concerning (ii), algorithms can be grouped into continuous- and discrete-time algorithms, where the latter constitute the majority in the literature on distributed optimization and continuous-time algorithms are rather used to address distributed control problems but also have regained interest in the last decade (Dürr, Zeng, & Ebenbauer, 2013; Feijer & Paganini, 2010; Gharesifard & Cortés, 2014; Niederländer & Cortés, 2015; Touri & Gharesifard, 2016; Wang & Elia, 2011). In this chapter, we focus on continuous-time algorithms. With respect to (i), there are two main approaches proposed in the literature:

a) Imposing structural constraints on the individual objective functions H_i and the constraints a, c. More precisely, it is assumed that H is additively separable, i.e., H_i is a function of z_i only, where $z_i \in \mathbb{R}$ is the ith component of z, and that the constraints are compatible with the network structure.

b) Using a *consensus-based* approach, i.e., the problem is divided into n smaller subproblems and an additional consensus constraint is introduced to enforce that all subproblems have a common solution.

While the first approach is appealing since it leads to subproblems with fewer variables, it is only applicable to a small class of problems or, vice versa, imposes high requirements on the network structure, and only few publications using this approach exist (Niederländer & Cortés, 2015). The consensus-based approach is much more common, and we will briefly explain it in the following. The core idea is to include a consensus constraint and write (1.1) equivalently as

$$
\begin{aligned}
\underset{\bar{z}\in\mathbb{R}^{np}}{\text{minimize}} \quad & \bar{H}(\bar{z}) = \sum_{i=1}^{n} H_i(\bar{z}_i) \\
\text{s.t.} \quad & a(\bar{z}_i) = 0 \\
& c(\bar{z}_i) \le 0 \\
& \bar{z}_i = \bar{z}_j \qquad \text{for all } i, j = 1, 2, \ldots, n,
\end{aligned}
\tag{2.2}
$$

where $\bar{z}_i \in \mathbb{R}^p$ can be interpreted as the ith agent's estimate of z. It is well-known that for strongly connected directed graphs with graph Laplacian $G \in \mathbb{R}^{n\times n}$ (see Appendix A.2 for a brief introduction to graph theory) the consensus constraint $\bar{z}_i = \bar{z}_j$ is equivalent to $(G \otimes I_p)\bar{z} = 0$ and hence can be enforced by constraints between neighboring agents only. The advantage of the latter formulation is that the optimization problem can now be split into n subproblems

$$
\begin{aligned}
\underset{\bar{z}_i\in\mathbb{R}^{p}}{\text{minimize}} \quad & H_i(\bar{z}_i) \\
\text{s.t.} \quad & a(\bar{z}_i) = 0 \\
& c(\bar{z}_i) \le 0 \\
& \bar{z}_i = \bar{z}_j \qquad \text{for all agents } j \text{ that are out-neighbors of agent } i,
\end{aligned}
\tag{2.3}
$$

$i = 1, 2, \dots, n$, which are coupled by the consensus constraint, hence requiring information exchange between the agents. However, this approach does neither reduce the dimension of the optimization variable nor the number of constraints of each agent. Since the couplings are directly related to the required network structure, much research has been done to reduce the required amount of coupling, or, from another perspective, to cope with these couplings under a given network structure from algorithmic side, leading to algorithms applicable, e.g., in the presence of time-varying and/or directed communication graphs. The majority of these approaches can be grouped into weighted-average-based approaches, push-sum-based approaches and primal-dual-based approaches, see Nedić and Liu (2018) and references therein.

Quite recently, a novel approach that does apply to but does not require a consensus-based reformulation and significantly reduces the requirements of approach a) has been proposed in Ebenbauer, Michalowsky, Grushkovskaya, and Gharesifard (2017) and extended in Michalowsky et al. (2017a, 2018); Michalowsky, Gharesifard, and Ebenbauer (2020). The following chapter builds upon these works and further generalizes the approach. The idea is to address the core problem of distributed optimization of limited local information by generating the missing information utilizing Lie bracket approximation techniques. More precisely, it is observed and exploited that, roughly speaking, Lie brackets correspond to additional edges in a communication graph (see Figure 2.1). This has already been studied to some degree in Belabbas (2013); Chen, Belabbas, and Başar (2015); Gharesifard (2017) regarding controllability questions of certain sparse control systems as well as in Costello and Egerstedt (2014) in terms of distributed computations. However, the implication that those Lie brackets give rise to distributed algorithms or controllers only utilizing local information has not been utilized.

2.1 Problem Formulation

Throughout this chapter, we consider optimization problems

$$\begin{aligned} \underset{z \in \mathbb{R}^p}{\text{minimize}} \quad & H(z) \\ \text{s.t.} \quad & a(z) = 0 \\ & c(z) \leq 0, \end{aligned} \tag{2.4}$$

where $H : \mathbb{R}^p \to \mathbb{R}$, $a : \mathbb{R}^p \to \mathbb{R}^{n_{\text{eq}}}$, $c : \mathbb{R}^p \to \mathbb{R}^{n_{\text{ineq}}}$, and $p, n_{\text{eq}}, n_{\text{ineq}} \in \mathbb{N}_{>0}$. We impose the following assumptions:

Assumption 1. The optimization problem (2.4) fulfills the following:

[A1] $H \in \mathcal{C}^2$ and strictly convex (see Definition 14).

[A2] For each $i \in \{1, \dots, n_{\text{eq}}\}$, the function $a_i \in \mathcal{C}^2$ is affine, and, for each $i \in \{1, \dots, n_{\text{ineq}}\}$, the function $c_i \in \mathcal{C}^2$ is convex, where a_i and c_i denote the ith component of a and c, respectively.

[A3] The minimum of (2.4) is attained. •

We note that **[A1]**, **[A2]** render problem (2.4) a convex optimization problem, hence, by **[A3]**, there is a unique minimizer which we denote by $z^\star$. Our primary goal is to design continuous-time optimization algorithms that converge to this minimizer in a sense that will be made clearer later and that can be implemented in a distributed fashion. More precisely, the latter means that we have a group of n agents available, each having access to limited local information only but capable of interchanging its information over a communication network. We thereby suppose that the number of agents is equal to the dimension of the optimization variable, i.e., $n = p$; we emphasize that this is not required for the following methodology to apply but leads to a more convenient notation. Throughout this chapter, we consider network structures that can be represented by a time-invariant directed graph $\mathcal{G} = (\mathcal{V}, \mathcal{E})$ consisting of a finite set of *nodes* denoted by $\mathcal{V} = \{1, 2, \dots, n\}$ and a finite set of *edges* denoted by $\mathcal{E} \subseteq \mathcal{V} \times \mathcal{V}$ that describes the connections between the nodes. Alternatively, such a graph $\mathcal{G}$ can be described by its graph Laplacian $G = [g_{ij}] \in \mathbb{R}^{n \times n}$; we refer the reader to Appendix A.2 for a brief overview of the required basics of graph theory. In the present setup, each node $i \in \mathcal{V}$ represents an agent and the terms node and agent will be used equivalently. We then associate to each agent $i \in \mathcal{V}$ a time-dependent state $\mathbf{x}_i(t) \in \mathbb{R}^{N_i}$, $N_i \in \mathbb{N}_{>0}$, consisting of the ith component of the optimization variable z_i and possibly other internal states required in the distributed algorithm to be designed. The edges then define the communication links between the agents in the following manner: if there is an edge from node i to node j, i.e., $(i, j) \in \mathcal{E}$, then agent i has access to the state of agent j and may hence utilize $\mathbf{x}_j$ in its dynamics (cf. Figure 1.2, (i)). Let

$$\mathcal{N}_{\mathcal{G}}(i) = \{i\} \cup \{j \in \mathcal{V} : (i, j) \in \mathcal{E}\} \tag{2.5}$$

denote the set of all out-neighbors of agent i as well as the agent itself. Using this definition, the vector of states agent i has access to is given by stacking all agent states $\mathbf{x}_k$ with $k \in \mathcal{N}_{\mathcal{G}}(i)$. In words, each agent has access to its own state as well as the states of its out-neighbors. In what follows, we denote this vector by $[\mathbf{x}_k]_{k \in \mathcal{N}_{\mathcal{G}}(i)}$ utilizing the notation explained in Appendix A.1.

We next clarify our understanding of a distributed algorithm. Suppose that, for each agent $i = 1, 2, \dots, n$, the agent dynamics are given as

$$\dot{\mathbf{x}}_i = \mathbf{f}_i(\mathbf{x}), \tag{2.6}$$

where $\mathbf{f}_i : \mathbb{R}^N \to \mathbb{R}^{N_i}$, and $\mathbf{x} = [\mathbf{x}_i]_{i=1,\dots,n} \in \mathbb{R}^N$, $N = \sum_{i=1}^n N_i$, is the stacked vector of all agent states. We then have the following definition of a distributed algorithm:

Definition 1 (Distributed Algorithm). We say that a continuous-time algorithm with agent dynamics of the form (2.6) and output $z = h(\mathbf{x})$ is *distributed w.r.t. the graph* $\mathcal{G} = (\mathcal{V}, \mathcal{E})$ if, for each $i \in \mathcal{V}$, there exists $\tilde{\mathbf{f}}_i$ such that $\mathbf{f}_i(\mathbf{x}) = \tilde{\mathbf{f}}_i([\mathbf{x}_k]_{k \in \mathcal{N}_{\mathcal{G}}(i)})$ for all $\mathbf{x} = [\mathbf{x}_i]_{i \in \mathcal{V}} \in \mathbb{R}^N$. •

In words, each agent may only use its own state as well as those of all its out-neighbors, thus imposing constraints on the stacked vector field $\mathbf{f} = [\mathbf{f}_1^\top, \mathbf{f}_2^\top, \dots, \mathbf{f}_n^\top]^\top$ or, equivalently, imposing sparsity constraints on the Jacobian of $\mathbf{f}$ that are in accordance with the graph

Laplacian G. The same definitions apply if $\mathbf{f}_i$ in (2.6) explicitly depends on time, with the additional requirement that the conditions must hold for all $t \in \mathbb{R}$. According to Definition 1, whether some algorithm (1.2) is distributed or not is independent of the output function h; therefore, throughout this chapter we mostly neglect that an algorithm also requires some output and only study the algorithm dynamics.

Consider now an algorithm with dynamics

$$\dot{x} = f(x), \tag{2.7}$$

where $x(t) = [x_i(t)]_{i=1,\ldots,N} \in \mathbb{R}^N$, $x_i(t) \in \mathbb{R}$. Suppose we want to implement (2.7) in a distributed fashion in the sense of Definition 1 or check whether this is possible at all. To this end, a preliminary step in which we assign the components x_i, $i = 1, 2, \ldots, N$, of the state to the n agents is required. We do not explicitly take this step into account here but assume such an assignment to be given. Note that we utilize boldface letters to distinguish the agent dynamics (2.6) from the dynamics of some algorithm (2.7). We emphasize that the pairs x, f and $\mathbf{x}, \mathbf{f}$ are equivalent up to reordering of the components and arguments (cf. Figure 2.4); we make this clearer later. The methodology we present will be applied to (2.7); still, a distributed implementation requires the agent dynamics (2.6). We then call (2.7) a distributed algorithm if the corresponding agent dynamics (2.6) are distributed in the sense of Definition 1. If (2.7) is distributed, we say that f is *admissible*. More generally, we can think of an admissible vector field f^{adm} as a vector field that yields distributed dynamics $\dot{x} = f^{\mathrm{adm}}(x)$ in the sense of Definition 1. We give a precise definition later, cf. Definition 3. A non-distributed algorithm (2.7) then contains admissible as well as non-admissible vector fields; in particular, the vector field f in (2.7) can be decomposed as

$$\dot{x} = f(x) = f^{\mathrm{adm}}(x) + f^{\neg\mathrm{adm}}(x), \tag{2.8}$$

where f^{adm} is admissible whereas $f^{\neg\mathrm{adm}}$ is not. In order to render the non-distributed algorithm (2.8) a distributed one, simply put, we need to render the non-admissible part admissible. While distributed implementations of various non-distributed algorithms have been derived, to the best of our knowledge, there exists no systematic way of determining a distributed algorithm that is capable of solving the same problem class as the non-distributed algorithm. In the following we aim to develop a methodology that allows to address this problem in quite general situations. The main idea is to make use of Lie bracket approximations. Lie bracket approximations have a long history in geometric nonlinear control (Brockett, 2014, and references therein), where they have been used to reveal and utilize hidden control directions of nonlinear systems, e.g., in motion planning problems for underactuated systems (Z. Li & Canny, 2012), but also in extremum seeking problems (Dürr, Stanković, Ebenbauer, & Johansson, 2013; Grushkovskaya, Zuyev, & Ebenbauer, 2018). Our approach further advances into that direction, where the hidden control direction is identified with a non-existing communication link. This idea is best illustrated by means of an example.

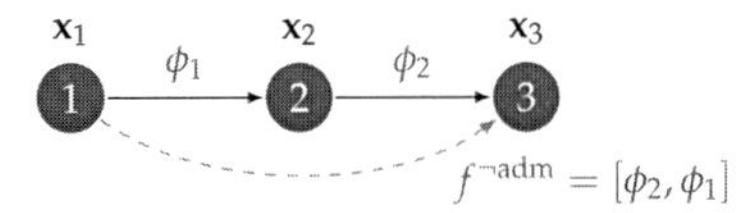

Figure 2.2. The communication graph from Example 1. The arrows indicate the directions of information sensing, i.e., agent 1 has access to the state of agent 2 and agent 2 has access to the state of agent 3. The dashed line depicts the fictitious edge created by Lie bracket approximations. We can associate the admissible vector fields ϕ_1, ϕ_2 to the edges and the non-admissible vector field $f^{\neg\mathrm{adm}}$ to the fictitious edge.

Example 1. Consider the dynamics

$$\begin{bmatrix} \dot{x}_1 \\ \dot{x}_2 \\ \dot{x}_3 \end{bmatrix} = \begin{bmatrix} x_2 + x_3 \\ -x_2 + 2x_3 \\ -x_3 \end{bmatrix} = f(x), \tag{2.9}$$

where $x_i(t) \in \mathbb{R}$, $i = 1, 2, 3$. Let a network of three agents with communication graph as depicted in Figure 2.2 be given. Suppose that the agent states are defined as $\mathbf{x}_i = x_i$, hence, to implement the dynamics (2.9), agent 1 and 2 both need access to the state of agent 2 and 3 and agent 3 only needs access to its own state. However, agent 1 does no have access to the state of agent 3 and we can decompose f in (2.9) into an admissible part f^{adm} and a non-admissible part $f^{\neg\mathrm{adm}}$ as $f(x) = f^{\mathrm{adm}}(x) + f^{\neg\mathrm{adm}}(x)$ with

$$f^{\mathrm{adm}}(x) = \begin{bmatrix} x_2 \\ -x_2 + 2x_3 \\ -x_3 \end{bmatrix}, \qquad f^{\neg\mathrm{adm}}(x) = \begin{bmatrix} x_3 \\ 0 \\ 0 \end{bmatrix}. \tag{2.10}$$

Thus, the algorithm (2.9) is not distributed in the sense of Definition 1. However, calculating the Lie bracket (see Appendix A.3 for a definition of Lie brackets) of two smartly chosen admissible vector fields $\phi_1(x) = e_1 x_2, \phi_2(x) = e_2 x_3$, with $e_i \in \mathbb{R}^3$ the ith unit vector, we observe that

$$\begin{aligned} [\phi_2, \phi_1](x) &= \frac{d\phi_1}{dx}(x)\phi_2(x) - \frac{d\phi_2}{dx}(x)\phi_1(x) \\ &= e_1 e_2^\top e_2 x_3 - e_2 e_3^\top e_1 x_2 \\ &= e_1 x_3 = f^{\neg\mathrm{adm}}(x). \end{aligned} \tag{2.11}$$

Thus, we have managed to write a non-admissible vector field in terms of a Lie bracket of admissible vector fields; the dynamics (2.9) can hence be written equivalently as

$$\dot{x} = f^{\mathrm{adm}}(x) + [\phi_2, \phi_1](x), \tag{2.12}$$

and the right-hand side consists of admissible vector fields and Lie brackets thereof only. It is well-known (Kurzweil & Jarník, 1987) from geometric control theory that the trajectories of the dynamics (2.12) can be "approximated arbitrarily close" by the trajectories of

$$\dot{x}^\sigma = f^{\mathrm{adm}}(x^\sigma) + U_1^\sigma(t)\phi_1(x^\sigma) + U_2^\sigma(t)\phi_2(x^\sigma), \tag{2.13}$$

given that $U_1^\sigma, U_2^\sigma : \mathbb{R} \to \mathbb{R}$ are chosen appropriately, e.g., $U_1^\sigma(t) = \sqrt{2\sigma}\sin(\sigma t)$ and $U_2^\sigma(t) = \sqrt{2\sigma}\cos(\sigma t)$, $\sigma > 0$. Then, given $x(0) = x^\sigma(0)$, the trajectories of (2.13) converge to those of (2.12) as σ increases in a sense that will be made clear later. Note that the right-hand side of (2.13) only consists of admissible vector fields and hence represents a distributed approximation of (2.12). Vividly, we can interpret this as having created a fictitious edge from node 1 to node 3 as indicated by the dashed line in Figure 2.2. •

The latter example captures the main idea of our approach: we first rewrite the non-admissible vector field $f^{\neg\mathrm{adm}}$ in a non-distributed algorithm (2.8) in terms of Lie brackets of admissible vector fields; Lie bracket approximation techniques then allow us to derive distributed approximations thereof. More precisely, our procedure consists of the following two steps (see Figure 2.3 for a schematic overview):

Step 1 Rewrite the non-admissible vector field $f^{\neg\mathrm{adm}}$ in terms of Lie brackets of admissible vector fields $\{\phi_1, \phi_2, \ldots, \phi_M\}$, i.e., write the algorithm dynamics (2.8) as

$$\dot{x} = f^{\mathrm{adm}}(x) + \sum_{B \in \mathcal{B}} v_B B(x), \tag{2.14}$$

where $\mathcal{B}$ is a set of Lie brackets of the admissible vector fields $\{\phi_1, \phi_2, \ldots, \phi_M\}$ and $v_B \in \mathbb{R}$.

Step 2 Derive distributed approximations of the non-distributed dynamics (2.14), i.e., design suitable functions U_k^σ, $k = 1, 2, \ldots, M$, parametrized by $\sigma > 0$ such that the trajectories of

$$\dot{x}^\sigma = f^{\mathrm{adm}}(x^\sigma) + \sum_{k=1}^{M} \phi_i(x^\sigma) U_k^\sigma(t), \tag{2.15}$$

converge to those of (2.14) as σ increases in a sense that will be made clear later.

Example 1 has already hinted how Step 1 and Step 2 can be addressed; the main subject of the remainder of this chapter is to generalize these ideas. We discuss Step 1 in Section 2.3 in detail and Step 2 in Section 2.4. Our methodology requires some non-distributed optimization algorithm that is able to solve (2.4) and from which we determine a distributed algorithm, see Figure 2.3. While, in principle, the procedure applies to a wide range of continuous-time algorithms (2.6), we concentrate on saddle-point dynamics as a prototype for a continuous-time optimization algorithm here, which we will introduce in the following section.

2.2 Saddle-Point Dynamics

In the sequel we briefly introduce saddle-point dynamics as a prototype of an algorithm able to solve an optimization problem (2.4). We emphasize that a detailed analysis of saddle-point dynamics is not within the scope of this thesis and a subject on its own. Saddle-point dynamics are a class of continuous-time dynamics with the property that they converge to a

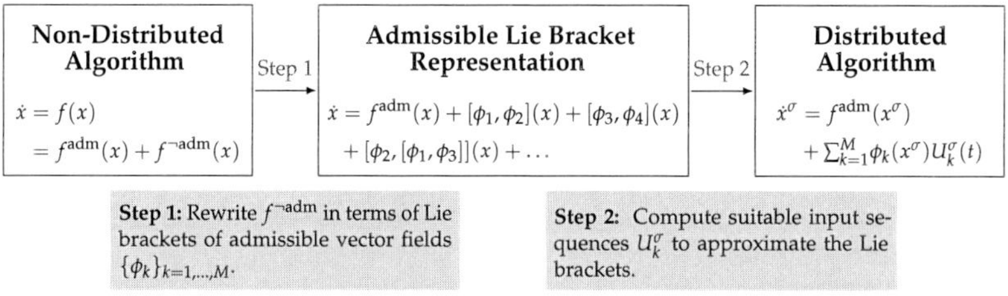

Non-Distributed Algorithm

$\dot{x} = f(x)$
$= f^{\text{adm}}(x) + f^{\neg\text{adm}}(x)$

Step 1

Admissible Lie Bracket Representation

$\dot{x} = f^{\text{adm}}(x) + [\phi_1, \phi_2](x) + [\phi_3, \phi_4](x)$
$+ [\phi_2, [\phi_1, \phi_3]](x) + \dots$

Step 2

Distributed Algorithm

$\dot{x}^{\sigma} = f^{\text{adm}}(x^{\sigma})$
$+ \sum_{k=1}^{M} \phi_k(x^{\sigma}) U_k^{\sigma}(t)$

Step 1: Rewrite $f^{\neg\text{adm}}$ in terms of Lie brackets of admissible vector fields $\{\phi_k\}_{k=1,\dots,M}$.

Step 2: Compute suitable input sequences U_k^{σ} to approximate the Lie brackets.

Figure 2.3. A schematic overview of the proposed two-step procedure: Starting from some non-distributed algorithm, we first rewrite the non-admissible vector field $f^{\neg\text{adm}}$ in terms of Lie brackets of admissible vector fields ϕ_k, $k = 1, \dots, M$, (Step 1, Section 2.3), and then determine a distributed approximation thereof utilizing Lie bracket approximation techniques (Step 2, Section 2.4).

saddle point of the Lagrangian associated to (2.4). Let $L : \mathbb{R}^p \times \mathbb{R}^{n_{\text{eq}}} \times \mathbb{R}_{\geq 0}^{n_{\text{ineq}}} \to \mathbb{R}$ denote the Lagrangian associated to (2.4), i.e.,

$$L(z, \nu, \lambda) = H(z) + \nu^\top a(z) + \lambda^\top c(z), \tag{2.16}$$

where $\nu \in \mathbb{R}^{n_{\text{eq}}}$, $\lambda \in \mathbb{R}_{\geq 0}^{n_{\text{ineq}}}$ are the associated Lagrange multipliers, also called *dual variables* in contrast to the *primal variable* z. We then have the following definition of a saddle point of the Lagrangian.

Definition 2 (Saddle Point). A point $(z^\star, \nu^\star, \lambda^\star) \in \mathbb{R}^p \times \mathbb{R}^{n_{\text{eq}}} \times \mathbb{R}_{\geq 0}^{n_{\text{ineq}}}$ is said to be a *saddle point* of L if, for all $z \in \mathbb{R}^p$, $\nu \in \mathbb{R}_{\text{eq}}^{n}$, $\lambda \in \mathbb{R}_{\geq 0}^{n_{\text{ineq}}}$, the inequalities

$$L(z^\star, \nu, \lambda) \leq L(z^\star, \nu^\star, \lambda^\star) \leq L(z, \nu^\star, \lambda^\star). \tag{2.17}$$

hold. •

It is well-known (Hiriart-Urruty & Lemaréchal, 2013, Corollary 4.4.2) that if the Lagrangian has a saddle point $(z^\star, \nu^\star, \lambda^\star)$, then $z^\star$ is a solution of (2.4). While **[A3]** ensures the existence of a minimizer $z^\star$, this does not guarantee the existence of a corresponding saddle point. Typically, so-called constraint qualifications are utilized to ensure the existence; here, we require the following assumption to hold:

Assumption 2 (Strong Slater Assumption). The constraints in (2.4) fulfill the *Strong Slater Assumption*, i.e., the gradients ∇a_i, $i \in \{1, 2, \dots, n_{\text{eq}}\}$ of the equality constraints are linearly independent and there exists a strictly feasible point $z_0 \in \mathbb{R}^p$, i.e., $a(z_0) = 0$ and $c(z_0) < 0$. •

This assumption ensures that the set of saddle points of the Lagrangian associated to (2.4) is non-empty, convex and compact, see Hiriart-Urruty and Lemaréchal (2013, Theorem 2.3.2). We denote by

$$\begin{aligned} \mathcal{M} := \Big\{ (z^\star, \nu^\star, \lambda^\star) \in \mathbb{R}^p \times \mathbb{R}^{n_{\text{eq}}} \times \mathbb{R}_{\geq 0}^{n_{\text{ineq}}} \mid L(z^\star, \nu, \lambda) \leq L(z^\star, \nu^\star, \lambda^\star) \leq L(z, \nu^\star, \lambda^\star) \\ \text{for all } z \in \mathbb{R}^p, \nu \in \mathbb{R}^{n_{\text{eq}}}, \lambda \in \mathbb{R}_{\geq 0}^{n_{\text{ineq}}} \Big\} \end{aligned} \tag{2.18}$$

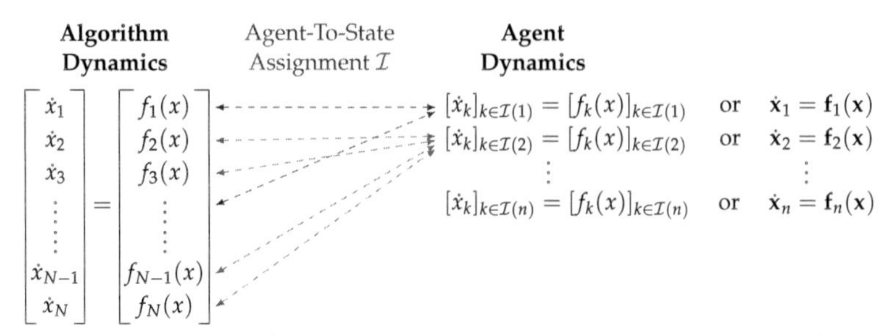

Figure 2.4. An illustration of the relation between the algorithm dynamics, the individual agent dynamics and the agent-to-state assignment that relates the components of these two dynamics. We emphasize that both representations are equivalent up to reordering.

the set of saddle points of the Lagrangian. We consider the following type of saddle-point dynamics from Dürr and Ebenbauer (2012) extended by equality constraints

$$\begin{aligned} \dot{z} &= -\nabla_z L(z,\nu,\lambda) &&= -\nabla H(z) - \tfrac{\partial a}{\partial z}(z)^\top \nu - \tfrac{\partial c}{\partial z}(z)^\top \lambda && \text{(2.19a)}\\ \dot{\nu} &= \nabla_\nu L(z,\nu,\lambda) &&= a(z) && \text{(2.19b)}\\ \dot{\lambda} &= \operatorname{diag}(\lambda)\nabla_\lambda L(z,\nu,\lambda) &&= \operatorname{diag}(\lambda)c(z). && \text{(2.19c)} \end{aligned}$$

One can show that any saddle point $(z^\star,\nu^\star,\lambda^\star) \in \mathcal{M}$ is an equilibrium of (2.19). The stability properties of these equilibria for (2.19) have been analyzed in Dürr (2015, Theorem 5.1.3) in detail in the case where no equality constraints are present. Since we are analyzing the stability of sets of equilibria instead of a singleton, appropriate stability definitions beyond the usual notion of asymptotic stability are required. Even more so, it turns out that we only have stability with respect to a restricted set of initial conditions, which requires a further refinement of the definitions. In Dürr (2015, Definition 2.1.6), the notion of *$\mathcal{I}$-global uniform asymptotic stability* is introduced, where $\mathcal{I}$ refers to the restricted set of initial conditions. We do not go into detail here, accept this for the moment and only use this notion for the statement of the following result; we give an interpretation afterwards.

Theorem 1. Consider (2.4) together with (2.19) and suppose that Assumption 1 and Assumption 2 hold. Let

$$\mathcal{R}(\mathcal{M}) = \Big\{(z,\nu,\lambda) \in \mathbb{R}^p \times \mathbb{R}^{n_{\mathrm{eq}}} \times \mathbb{R}^{n_{\mathrm{ineq}}}_{\geq 0} : \lambda \in \mathbb{R}^{n_{\mathrm{ineq}}}_{>0}\Big\}. \tag{2.20}$$

Then the set of saddle points $\mathcal{M}$ defined in (2.18) is $\mathcal{R}(\mathcal{M})$-globally uniformly asymptotically stable for (2.19). •

Given that Theorem 1 can be proven with minor modifications of the proof of Dürr (2015, Theorem 5.1.3), in Appendix B.1.1 we only provide the steps of the proof that are different. The essence to be taken from the stability result in Theorem 1 is as follows: for any initial condition $(z(0), \nu(0), \lambda(0)) \in \mathcal{R}(\mathcal{M})$, the solutions of (2.19) converge to the set of saddle points $\mathcal{M}$ and hence $z(t)$ in (2.19) converges to the solution of the optimization problem (2.4) as t tends to infinity. Thus, (2.19) indeed is an algorithm of the form (1.2) that is able to solve (2.4). More precisely, let

$$x(t) = \begin{bmatrix} z(t)^\top & \nu(t)^\top & \lambda(t)^\top \end{bmatrix}^\top \in \mathbb{R}^N, \tag{2.21}$$

$N = n + n_{\text{eq}} + n_{\text{ineq}}$, denote the *complete state* of (2.19). Then the dynamics (2.19) together with the output $h(x(t)) = z(t)$ is an optimization algorithm of the form (1.2) that solves (2.4). In the remainder of this chapter, we consider saddle-point dynamics as an exemplary non-distributed algorithm to apply our two-step procedure (see 2.3) to. To this end, as mentioned beforehand, it is first required to assign to each agent a subset of the components x_i, $i = 1, 2, \ldots, N$, of the complete state (2.21). More formally, we need to define a set-valued map $\mathcal{I} : \{1, 2, \ldots, n\} \rightrightarrows \{1, 2, \ldots, N\}$ that fulfills (i) the family of sets $\mathcal{I}(i)$, $i = 1, 2, \ldots, n$, is disjoint, and (ii) $\bigcup_{i=1}^{n} \mathcal{I}(i) = \{1, 2, \ldots, N\}$. In the following we call $\mathcal{I}$ an *agent-to-state assignment*. $\mathcal{I}(i)$ then returns the index set of components of x that are assigned to agent i and the agent's state $\mathbf{x}_i$, $i = 1, 2, \ldots, n$, is given by

$$\mathbf{x}_i = [x_k]_{k \in \mathcal{I}(i)}. \tag{2.22}$$

Figure 2.4 illustrates this assignment process; we emphasize once more that we utilize boldface letters $\mathbf{x}_i$ to distinguish from a real-valued component x_i of the complete state vector. It is understood that the choice of the agent-to-state assignment $\mathcal{I}$ is important since this determines how much and which information is allocated to each agent. We briefly discuss that in Section 2.5.1; from now on we assume such an agent-to-state assignment to be given by

$$\mathcal{I}(i) = \{i\} \cup \{j = n + k \mid k \in \mathcal{I}_{\text{eq}}(i)\} \cup \{j = n + n_{\text{eq}} + k \mid k \in \mathcal{I}_{\text{ineq}}(i)\}, \tag{2.23}$$

where $\mathcal{I}_{\text{eq}}(i) \subseteq \{1, 2, \ldots, n_{\text{eq}}\}$, $\mathcal{I}_{\text{ineq}}(i) \subseteq \{1, 2, \ldots, n_{\text{ineq}}\}$, $i = 1, 2, \ldots, n$, denote the disjoint index sets of equality and inequality constraints assigned to agent i. In a nutshell, this means that we assign to each agent i the component z_i of the primal variable as well as a subset of the constraints and hence the corresponding dual variables. We make this more apparent in the following example that we will reconsider several times to explain the different steps in the general procedure from Figure 2.3.

Example 2. Consider the graph of $n = 5$ nodes as depicted in Figure 2.5. The goal is to design a distributed algorithm for solving an optimization problem of the form (2.4), where $n_{\text{eq}} = 1$, $n_{\text{ineq}} = 3$, and

$$H(z) = \sum_{i=1}^{5} \tfrac{1}{2}(z_i - i)^2, \qquad a(z) = 2z_2 - z_5, \qquad c(z) = \begin{bmatrix} z_1^2 + z_2^2 - 4 \\ z_1 + z_3 - 2 \\ z_4^2 - z_4 z_1 + z_1^2 - 100 \end{bmatrix}. \tag{2.24}$$

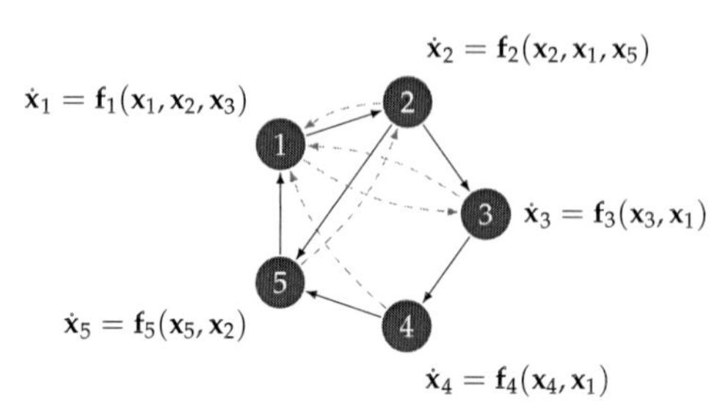

Figure 2.5. The communication graph corresponding to Example 2 together with the agent dynamics. The dashed lines indicate the edges missing to render the dynamics distributed. The presented methodology provides a way to create these missing links in a certain sense.

Note that Assumption 1 and Assumption 2 are fulfilled. The corresponding saddle-point dynamics (2.19) are then given by

$$\begin{bmatrix}\dot{z}_1\\\dot{z}_2\\\dot{z}_3\\\dot{z}_4\\\dot{z}_5\end{bmatrix} = \begin{bmatrix}z_1-1\\z_2-2\\z_3-3\\z_4-4\\z_5-5\end{bmatrix} + \begin{bmatrix}0\\2\\0\\0\\-1\end{bmatrix}\nu_1 + \begin{bmatrix}2z_1 & 1 & -z_4+2z_1\\2z_2 & 0 & 0\\0 & 1 & 0\\0 & 0 & 2z_4-z_1\\0 & 0 & 0\end{bmatrix}\begin{bmatrix}\lambda_1\\\lambda_2\\\lambda_3\end{bmatrix} \tag{2.25a}$$

$$\dot{\nu}_1 = 2z_2 - z_5 \tag{2.25b}$$

$$\begin{bmatrix}\dot{\lambda}_1\\\dot{\lambda}_2\\\dot{\lambda}_3\end{bmatrix} = \begin{bmatrix}\lambda_1(z_1^2+z_2^2-4)\\\lambda_2(z_1+z_3-2)\\\lambda_3(z_4^2-z_4z_1+z_1^2-100)\end{bmatrix}. \tag{2.25c}$$

With the complete state vector being defined as

$$x(t) = \begin{bmatrix}z_1(t) & z_2(t) & z_3(t) & z_4(t) & z_5(t) & \nu_1(t) & \lambda_1(t) & \lambda_2(t) & \lambda_3(t)\end{bmatrix}^\top \in \mathbb{R}^9, \tag{2.26}$$

we can write (2.25) equivalently as

$$\dot{x} = f(x) = \begin{bmatrix}x_1 - 1 + 2x_1x_7 + x_8 + x_9(2x_1 - x_4)\\ x_2 - 2 + 2x_6 + 2x_2x_7\\ x_3 - 3 + x_8\\ x_4 - 4 + 2x_4x_9 - x_1x_9\\ x_5 - 5 - x_6\\ 2x_2 - x_5\\ x_7(x_1^2 + x_2^2 - 4)\\ x_8(x_3 - 2)x_8x_1\\ x_9(x_4^2 - 100) + x_9(x_1^2 - x_1x_4)\end{bmatrix}. \tag{2.27}$$

We assume the agent-to-state assignment to be given by

$$\mathcal{I}(1) = \{1,7\} \qquad \mathcal{I}(2) = \{2,6\} \qquad \mathcal{I}(3) = \{3,8\} \tag{2.28a}$$
$$\mathcal{I}(4) = \{4,9\} \qquad \mathcal{I}(5) = \{5\} \tag{2.28b}$$

resulting in the agent states

$$\mathbf{x}_1 = \begin{bmatrix} x_1 & x_7 \end{bmatrix} = \begin{bmatrix} z_1 & \lambda_1 \end{bmatrix}, \qquad \mathbf{x}_3 = \begin{bmatrix} x_3 & x_8 \end{bmatrix} = \begin{bmatrix} z_3 & \lambda_2 \end{bmatrix}, \qquad \mathbf{x}_5 = x_5 = z_5, \tag{2.29a}$$
$$\mathbf{x}_2 = \begin{bmatrix} x_2 & x_6 \end{bmatrix} = \begin{bmatrix} z_2 & \nu_1 \end{bmatrix}, \qquad \mathbf{x}_4 = \begin{bmatrix} x_4 & x_9 \end{bmatrix} = \begin{bmatrix} z_4 & \lambda_3 \end{bmatrix}. \tag{2.29b}$$

For example, this means that the first inequality constraint c_1, and hence the corresponding dual variable λ_1, is assigned to agent 1. For the agent dynamics we then obtain

$$\dot{\mathbf{x}}_1 = \mathbf{f}_1(\mathbf{x}_1, \mathbf{x}_2, \mathbf{x}_3) = \begin{bmatrix} x_1 - 1 + 2x_1x_7 + x_8 + x_9(2x_1 - x_4) \\ x_7(x_1^2 + x_2^2 - 4) \end{bmatrix} \tag{2.30a}$$
$$\dot{\mathbf{x}}_2 = \mathbf{f}_2(\mathbf{x}_2, \mathbf{x}_1, \mathbf{x}_5) = \begin{bmatrix} x_2 - 2 + 2x_6 + 2x_2x_7 \\ 2x_2 - x_5 \end{bmatrix} \tag{2.30b}$$
$$\dot{\mathbf{x}}_3 = \mathbf{f}_3(\mathbf{x}_3, \mathbf{x}_1) \quad = \begin{bmatrix} x_3 - 3 + x_8 \\ x_8(x_3 - 2)x_8x_1 \end{bmatrix} \tag{2.30c}$$
$$\dot{\mathbf{x}}_4 = \mathbf{f}_4(\mathbf{x}_4, \mathbf{x}_1) \quad = \begin{bmatrix} x_4 - 4 + 2x_4x_9 - x_1x_9 \\ x_9(x_4^2 - 100) + x_9(x_1^2 - x_1x_4) \end{bmatrix} \tag{2.30d}$$
$$\dot{\mathbf{x}}_5 = \mathbf{f}_5(\mathbf{x}_5, \mathbf{x}_2) \quad = x_5 - 5 - x_6. \tag{2.30e}$$

Thus, the dynamics (2.30) are not distributed according to Definition 1. Indeed, it requires the additional communication links depicted by dashed arrows in Figure 2.5 to render the dynamics distributed. The methodology that we present can be interpreted as providing these missing links by following the two-step procedure from Figure 2.3. •

As also demonstrated in the latter example, although saddle-point dynamics are known to yield distributed dynamics in the sense of Definition 1 in some situations, this is not the case in general. Indeed, even for very basic consensus problems, saddle-point dynamics do not yield distributed implementations if the communication graph is directed (Gharesifard & Cortés, 2014). In the next two sections, we apply the two-step procedure sketched in Section 2.1 to obtain distributed dynamics that approximate the non-distributed saddle-point dynamics (2.19). Therein, the saddle-point dynamics (2.19) play the role of (2.8) (viz. (2.14)); following Step 1, we hence need to represent the non-admissible vector field in (2.19) in terms of Lie brackets of admissible vector fields. This is the main subject of the following section.

2.3 Admissible Lie Bracket Representations

In this section, we discuss Step 1 in the proposed procedure of rewriting non-admissible vector fields in terms of Lie brackets of admissible vector fields. We have already seen in Example 1 that it is indeed possible to rewrite non-admissible vector fields in terms of Lie brackets of admissible vector fields. In the remainder we discuss in detail how to generalize this idea. To this end, we first give a proper definition of an admissible vector field.

Definition 3 (Admissible Vector Field). Consider a graph $\mathcal{G}$ of n nodes $1, 2, \ldots, n$ together with an agent-to-state assignment $\mathcal{I} : \{1, 2, \ldots, n\} \rightrightarrows \{1, 2, \ldots, N\}$. Let $\mathbf{x}_i$ defined by (2.22) denote the state of agent i, $i = 1, 2, \ldots, n$. We say that a vector field $f : \mathbb{R}^N \to \mathbb{R}^N$ with components $f_k : \mathbb{R}^N \to \mathbb{R}$, $k = 1, 2, \ldots, N$, is *admissible* for the graph $\mathcal{G}$ w.r.t. the agent-to-state assignment $\mathcal{I}$ if, for each $k \in \mathcal{I}(i)$, $i = 1, 2, \ldots, n$, there exists a real-valued function $\tilde{f}_k$ such that

$$f_k(x) = \tilde{f}_k([\mathbf{x}_j]_{j \in \mathcal{N}_{\mathcal{G}}(i)}) \tag{2.31}$$

for all $x \in \mathbb{R}^N$. We call the vector field f *non-admissible* if such functions do not exist. •

In the following we often only speak of admissible vector fields neglecting the dependence on the graph as well as the agent-to-state assignment. Reconsidering Definition 1, any admissible vector field $f : \mathbb{R}^N \to \mathbb{R}^N$ then yields distributed agent dynamics (2.6) with

$$\mathbf{f}_i(\mathbf{x}) = [\tilde{f}_k([\mathbf{x}_j]_{j \in \mathcal{N}_{\mathcal{G}}(i)})]_{k \in \mathcal{I}(i)} \tag{2.32}$$

In this sense, we can think of an admissible vector field as a vector field that can be implemented in a distributed fashion. We further denote by

$$\Phi_{\text{all}}^{\text{adm}}(\mathcal{G}, \mathcal{I}) = \{\phi : \mathbb{R}^N \to \mathbb{R}^N \mid \phi \text{ is admissible for } \mathcal{G} \text{ w.r.t. } \mathcal{I}\} \tag{2.33}$$

the set of all admissible vector fields for the graph $\mathcal{G}$ w.r.t. the agent-to-state assignment $\mathcal{I}$. In view of the proposed two-step procedure, we aim to rewrite the non-admissible vector field $f^{\neg\text{adm}}$ in a non-distributed algorithm (2.8) in terms of Lie brackets of admissible vector fields. To this end, typically only a finite subset of all admissible vector fields is required. We denote this set of required admissible vector fields by $\Phi^{\text{adm}}(\mathcal{G}, \mathcal{I}) \subset \Phi_{\text{all}}^{\text{adm}}(\mathcal{G}, \mathcal{I})$ in the following. We then call a representation of a non-admissible vector field by Lie brackets of admissible vector fields an *admissible Lie bracket representation*; we give a precise definition in the following.

Definition 4 (Admissible Lie Bracket Representation). Consider a graph $\mathcal{G}$ of n nodes $1, 2, \ldots, n$ and let an agent-to-state assignment $\mathcal{I} : \{1, 2, \ldots, n\} \rightrightarrows \{1, 2, \ldots, N\}$ be given. We say that a vector field $f : \mathbb{R}^N \to \mathbb{R}^N$ *has an admissible Lie bracket representation* w.r.t. $(\mathcal{G}, \mathcal{I})$ if there exist Lie brackets $B_i \in \mathcal{LB}r(\Phi_{\text{all}}^{\text{adm}}(\mathcal{G}, \mathcal{I}))$, $i = 1, 2, \ldots, m$, $m \in \mathbb{N}_{>0}$ arbitrary, and corresponding coefficients $v_{B_i} \in \mathbb{R}$ such that

$$f(x) = \sum_{i=1}^{m} v_{B_i} B_i(x) \tag{2.34}$$

for all $x \in \mathbb{R}^N$. We further say that a dynamic system $\dot{x} = f(x)$ has an admissible Lie bracket representation w.r.t. $(\mathcal{G}, \mathcal{I})$ if the vector field f has an admissible Lie bracket representation w.r.t. $(\mathcal{G}, \mathcal{I})$. •

Similar as beforehand, in the following we often neglect the dependence on the graph as well as the agent-to-state assignment. Simply put, a vector fields has an admissible Lie bracket representation if it can be rewritten as a (weighted) sum of Lie brackets of admissible vector fields. As an example, suppose that $\phi_1, \phi_2, \phi_3 \in \Phi_{\text{all}}^{\text{adm}}(\mathcal{G}, \mathcal{I})$ and let

$$f^{\neg\text{adm}}(x) = B(x), \qquad B = \Big[\big[[\phi_1, \phi_2], \phi_3 \big], [\phi_2, \phi_3] \Big]. \tag{2.35}$$

We then say that $f^{\neg\text{adm}}$ has an admissible Lie bracket representation and call B an admissible Lie bracket representation of $f^{\neg\text{adm}}$.

Following the two-step procedure, given non-distributed dynamics (2.8), our goal is then to find admissible Lie bracket representations for the non-admissible vector field $f^{\neg\text{adm}}$ appearing therein. To this end, it is more convenient to decompose the vector field $f^{\neg\text{adm}} : \mathbb{R}^N \to \mathbb{R}^N$ into its components as

$$f^{\neg\text{adm}}(x) = \sum_{k=1}^{N} e_k f_k^{\neg\text{adm}}(x), \tag{2.36}$$

where $e_k \in \mathbb{R}^N$ is the kth unit vector, and treat every single non-admissible vector field $e_k f_k^{\neg\text{adm}}$ separately. While we are particularly interested in deriving admissible Lie bracket representations of all non-admissible vector fields of this form appearing in the saddle-point dynamics (2.19), in Section 2.3.1 and Section 2.3.2 we first address the following more general question: Which classes of non-admissible vector fields have an admissible Lie bracket representation? To this end, we go the other way round: first, in Section 2.3.1, we consider a subclass of admissible vector fields and study which non-admissible vector fields can be computed by taking Lie brackets thereof. This approach was also followed in Michalowsky et al. (2018); we generalize these ideas here considering a larger class of Lie brackets. In Section 2.3.2, we then study smart choices of the admissible vector fields yielding important special cases of the admissible Lie bracket representations derived in Section 2.3.1. These special cases can be seen as building blocks for systematically finding admissible Lie bracket representations of non-admissible vector fields; this is illustrated in Section 2.3.3 where we apply the results to the specific dynamics (2.19).

2.3.1 General Case

In Example 1 we have considered Lie brackets of two smartly chosen admissible vector fields yielding a non-admissible vector field. In this section we aim to generalize this by considering Lie brackets of arbitrary degree of a class of admissible vector fields. In particular, we study Lie brackets that are, simply put, "built along a path". To motivate and explain this idea, we reconsider Example 1.

Example 1 (continuing from p. 14). Let the graph be extended by a fourth node with associated real-valued state $\mathbf{x}_4 = x_4$ as depicted in Figure 2.6 and let the dynamics be given by $\dot{x}_4 = -x_4$. Suppose now that agent 1 wants to use x_4 instead of x_3 in its dynamics, i.e., the

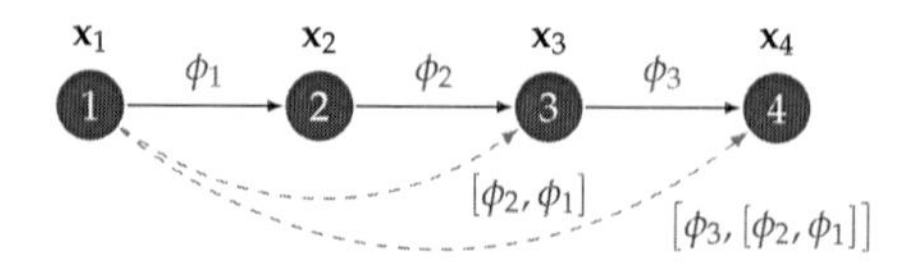

Figure 2.6. The extended communication graph considered in Example 1 (continued). We associate three admissible vector fields ϕ_1, ϕ_2, ϕ_3 to the edges and take Lie brackets thereof. Two particular Lie brackets can be associated to fictitious edges indicated by the dashed lines.

non-admissible part in the algorithm dynamics is now given by $f^{\neg\mathrm{adm}}(x) = e_1 x_4$, $e_1 \in \mathbb{R}^4$. If we take the fictitious edge from node 1 to node 3 with associated non-admissible vector field $e_1 x_3 = [\phi_2, \phi_1]$ to be given, then the situation is basically the same as in Example 1. Motivated by this observation, we let $\phi_3(x) = e_3 x_4$, which is an admissible vector field that can be associated to the edge from agent 3 to agent 4. Similar to (2.11), we then compute

$$[\phi_3, [\phi_2, \phi_1]](x) = e_1 x_4 = f^{\neg\mathrm{adm}}(x). \tag{2.37}$$

Hence, we have determined an admissible Lie bracket representation $B = [\phi_3, [\phi_2, \phi_1]]$ of the non-admissible vector field $f^{\neg\mathrm{adm}}$. We can associate this admissible Lie bracket representation to a fictitious edge from node 1 to node 4, see Figure 2.6 •

The latter example suggests that we could go on like that when the graph is extended in the same manner by additional nodes. This is indeed possible; still, the question arises how we can *systematically* choose the admissible vector fields that constitute the admissible Lie bracket representation. We address this question in the following. Motivated by Example 1 (continued), the idea of what follows is to consider a path in some graph $\mathcal{G}$, associate to each edge along the path an admissible vector field, and then study Lie brackets thereof. To keep the already cumbersome notation on an acceptable level, we assume in this section that the path is given as depicted in Figure 2.7. We emphasize that the same results hold for arbitrary paths replacing the $\mathfrak{p}_{1,r}$ by a path $\mathfrak{p}_{i_1,i_r} = \langle i_1 \mid i_2 \mid \ldots \mid i_r \rangle$.

Consider the path $\mathfrak{p}_{1,r}$ depicted in Figure 2.7. As mentioned beforehand, our approach is to associate to each edge along the path an admissible vector field and study Lie brackets thereof. In particular, we associate to each edge $\langle k \mid k+1 \rangle$, $k \in \{1, 2, \ldots, r-1\}$, a vector field

$$h_{k,j}(x) = e_j F_j(\mathbf{x}_k, \mathbf{x}_{k+1}), \qquad j \in \mathcal{I}(k), \tag{2.38}$$

where F_j is a smooth real-valued function and $e_j \in \mathbb{R}^N$ is the jth unit vector. Note that $h_{k,j}$ is admissible since $j \in \mathcal{I}(k)$. In the following we denote by $j_k \in \mathcal{I}(k)$ a specific choice of j in (2.38) and let

$$\psi : \langle k \mid k+1 \rangle \mapsto j_k, \qquad k \in \{1, 2, \ldots, \ell(\mathfrak{p}_{1,r})\}, \tag{2.39}$$

denote the map that associates to each edge $\langle k \mid k+1 \rangle$ the corresponding index $j_k \in \mathcal{I}(k)$. We mostly use j_k instead of $\psi(\langle k \mid k+1 \rangle)$ in the favor of a shorter notation.

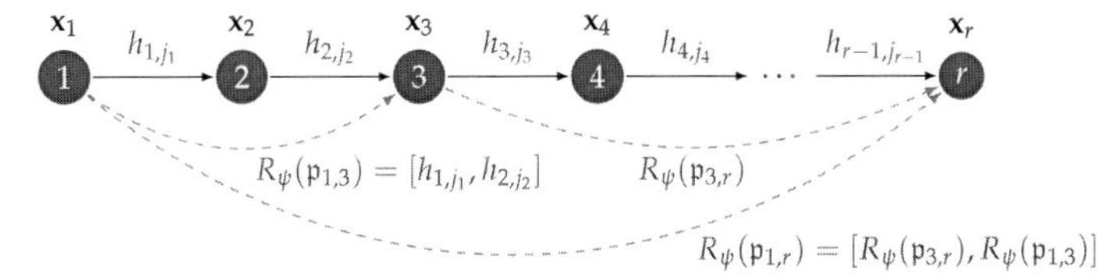

Figure 2.7. A path $\mathfrak{p}_{1,r}$, where we associate to the kth edge along the path the admissible vector field $h^k_{j_k}$, $j_k \in \mathcal{I}(k)$, and three exemplary admissible Lie bracket representations in terms of R_ψ of non-admissible vector fields with associated fictitious edges depicted by the dashed lines.

Remark 1. We note that (2.38) is not the most general form of admissible vector fields. In fact, F_j can be a function of all states of all out-neighboring agents of agent k. This generalization allows for more general functions to be represented as Lie brackets of admissible vector fields. Still, as it turns out in the following, the class of vector fields (2.38) is rich enough to find admissible Lie bracket representations of a large class of non-admissible vector fields. •

We now consider Lie brackets of admissible vector fields of the form (2.38) that are, in rough words, built along the path $\mathfrak{p}_{1,r}$ depicted in Figure 2.7. To this end, for a map ψ defined in (2.39) corresponding to a path $\mathfrak{p}_{1,r}$, we recursively define a mapping R_ψ from a subpath $\mathfrak{p}$ of $\mathfrak{p}_{1,r}$ to the set of vector fields on $\mathbb{R}^N$ as follows:

- for $\ell(\mathfrak{p}) = 1$, we define

$$R_\psi(\mathfrak{p}) = h_{k,j_k}, \qquad k = \text{head}(\mathfrak{p}), j_k = \psi(\mathfrak{p}); \tag{2.40}$$

- for $\ell(\mathfrak{p}) \geq 2$, we define

$$R_\psi(\mathfrak{p}) = \big[R_\psi(\mathfrak{q}^c), R_\psi(\mathfrak{q})\big], \tag{2.41}$$

  where $\mathfrak{q} \in \text{subpath}_{i\bullet}(\mathfrak{p})$, $i \in \text{head}(\mathfrak{p})$, is any subpath of $\mathfrak{p}$.

Since $\psi(\mathfrak{p}) \in \mathcal{I}(\text{head}(\mathfrak{p}))$ for $\ell(\mathfrak{p}) = 1$ by definition, we infer that h_{k,j_k} in (2.40) is admissible; hence $R_\psi(\mathfrak{p})$ is a Lie bracket of admissible vector fields for any $\mathfrak{p}$. We note that $R_\psi(\mathfrak{p})$ also depends on the choice of the functions F_j which is not reflected in the notation. We accept this ambiguity here to avoid an even more cumbersome notation and assume that this is understood. We can picture R_ψ as a mapping that returns "arbitrarily structured Lie brackets along a path" built from admissible vector fields (2.38); the choice of subpaths in (2.41) then determines the structure. This interpretation will become clearer after the following result which we prove in Appendix B.1.2.

Lemma 1. Consider a graph $\mathcal{G}$ of n nodes $1, 2, \ldots, n$ together with an agent-to-state assignment $\mathcal{I} : \{1, 2, \ldots, n\} \rightrightarrows \{1, 2, \ldots, N\}$. Let $\mathfrak{p}_{1,r}$ as depicted in Figure 2.7 be a simple path in $\mathcal{G}$. Then, for any $l \in \{1, 2, \ldots, r-1\}$ and any map $\psi : \langle k \mid k+1 \rangle \mapsto j_k$, $j_k \in \mathcal{I}(k)$, $k = 1, 2, \ldots, \ell(\mathfrak{p}_{1,r})$, we have

$$R_{\psi}(\mathfrak{p}_{l,r})(x) = e_{j_l} \tilde{F}_{l,r}(x) \tag{2.42}$$

with $e_{j_l} \in \mathbb{R}^N$ and

$$\tilde{F}_{l,r}(x) = F_{j_{r-1}}(\mathbf{x}_{r-1}, \mathbf{x}_r) \prod_{k=l}^{r-2} \frac{\partial F_{j_k}}{\partial x_{j_{k+1}}}(\mathbf{x}_k, \mathbf{x}_{k+1}), \tag{2.43}$$

and $R_{\psi}(\mathfrak{p}_{l,r})$ is a Lie bracket of admissible vector fields. •

By the latter Lemma, each non-admissible vector field that takes the same form as the right-hand side of (2.42) has an admissible Lie bracket representation. In Section 2.3.2 we discuss particular choices of the functions F_{j_k} that lead to admissible Lie bracket representations of non-admissible vector fields that occur in the problem at hand. It is worth mentioning that Lemma 1 does not classify the whole set of vector fields having an admissible Lie bracket representation since we limited ourselves to a single path as well as a subclass of admissible vector fields, see also Remark 1. We also note that admissible Lie bracket representations are highly non-unique; in particular, the choice of subpaths in (2.41) is arbitrary and results in different bracket structures representing the same non-admissible vector fields. We come back to that later in Section 2.4.1 and utilize this degree of freedom to simplify the construction of the approximating inputs. As an example, let $r = 4$. Then, according to Lemma 1, the following five brackets are different admissible Lie bracket representations of the same vector field and correspond to different choices of the subpaths (illustrated on the right-hand side):

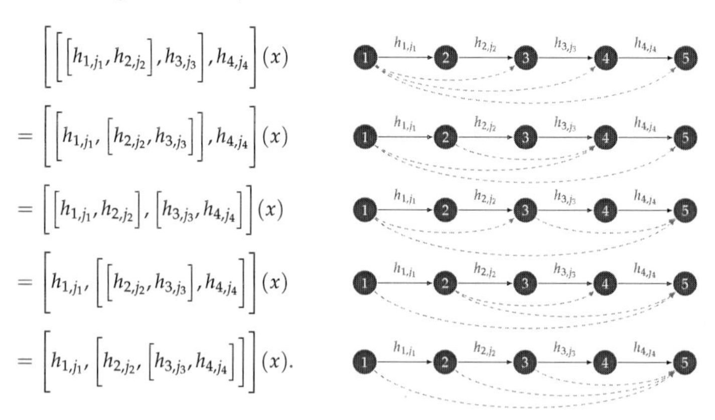

$$\begin{aligned}
&\Big[\big[[h_{1,j_1}, h_{2,j_2}], h_{3,j_3}\big], h_{4,j_4}\Big](x)\\
=&\Big[\big[h_{1,j_1}, [h_{2,j_2}, h_{3,j_3}]\big], h_{4,j_4}\Big](x)\\
=&\Big[[h_{1,j_1}, h_{2,j_2}], [h_{3,j_3}, h_{4,j_4}]\Big](x)\\
=&\Big[h_{1,j_1}, \big[[h_{2,j_2}, h_{3,j_3}], h_{4,j_4}\big]\Big](x)\\
=&\Big[h_{1,j_1}, \big[h_{2,j_2}, [h_{3,j_3}, h_{4,j_4}]\big]\Big](x).
\end{aligned}$$

In other words, on the set of vector fields considered here, the Lie bracket is an associative operation in the sense as explained above. We note that the number of different brackets is given by the Catalan numbers (Stanley, 2015).

Remark 2. We briefly discuss the implications of the latter observation. To this end, let $r = 3$. Then, according to Lemma 1, the following two brackets are different admissible Lie bracket representations of the same vector field:

$$\left[h_{1,j_1}, \left[h_{2,j_2}, h_{3,j_3}\right]\right](x) = \left[\left[h_{1,j_1}, h_{2,j_2}\right], h_{3,j_3}\right](x). \tag{2.44}$$

By the Jacobi identity (A.7), we infer that

$$\left[h_{1,j_1}, \left[h_{2,j_2}, h_{3,j_3}\right]\right](x) + \left[h_{2,j_2}, \left[h_{3,j_3}, h_{1,j_1}\right]\right](x) + \left[h_{3,j_3}, \left[h_{1,j_1}, h_{2,j_2}\right]\right](x) = 0 \tag{2.45}$$

for all $x \in \mathbb{R}^N$. Together with (2.44) and the skew-symmetry property (A.6), this implies that

$$\left[h_{2,j_2}, \left[h_{3,j_3}, h_{1,j_1}\right]\right](x) = 0 \tag{2.46}$$

for all $x \in \mathbb{R}^N$. We later exploit this fact to simplify the construction of approximating inputs as required for Step 2 in the procedure from Figure 2.3. •

2.3.2 Special Cases

In this section we discuss particular choices of the functions F_{j_k} in Lemma 1 that lead to admissible Lie bracket representations of non-admissible vector fields that occur in the problem at hand. In order to give an insight into what type of non-admissible vector fields appear in the particular application, we next reconsider Example 2.

Example 2 (continuing from p. 19). Reconsider the saddle-point dynamics (2.27). We first observe that

$$\dot{x} = \begin{bmatrix} x_1 - 1 + 2x_1x_7 \\ x_2 - 2 + 2x_6 \\ x_3 - 3 + x_8 \\ x_4 - 4 + 2x_4x_9 \\ x_5 - 5 \\ 2x_2 - x_5 \\ x_7(x_1^2 + x_2^2 - 4) \\ x_8(x_3 - 2) \\ x_9(x_4^2 - 100) \end{bmatrix} + \begin{bmatrix} x_8 + x_9(2x_1 - x_4) \\ 2x_2x_7 \\ 0 \\ -x_1x_9 \\ -x_6 \\ 0 \\ 0 \\ x_8x_1 \\ x_9(x_1^2 - x_1x_4) \end{bmatrix} = f^{\text{adm}}(x) + f^{\neg\text{adm}}(x), \tag{2.47}$$

where the first vector field f^{adm} is admissible whereas $f^{\neg\mathrm{adm}}$ is not. Decomposing the non-admissible vector field into its components and the components summands, we then have

$$f^{\neg\mathrm{adm}}(x) = \sum_{i=1}^{9} f_i^{\neg\mathrm{adm}}(x), \tag{2.48}$$

where

$$f_1^{\neg\mathrm{adm}}(x) = e_1 x_8, \qquad f_2^{\neg\mathrm{adm}}(x) = e_1 2 x_9 x_1, \qquad f_3^{\neg\mathrm{adm}}(x) = -e_1 x_9 x_4, \tag{2.49a}$$
$$f_4^{\neg\mathrm{adm}}(x) = e_2 2 x_2 x_7, \qquad f_5^{\neg\mathrm{adm}}(x) = -e_4 x_1 x_9, \qquad f_6^{\neg\mathrm{adm}}(x) = -e_5 x_6, \tag{2.49b}$$
$$f_7^{\neg\mathrm{adm}}(x) = e_8 x_8 x_1, \qquad f_8^{\neg\mathrm{adm}}(x) = e_9 x_9 x_1^2, \qquad f_9^{\neg\mathrm{adm}}(x) = -e_9 x_9 x_1 x_4. \tag{2.49c}$$

Our goal is then to find admissible Lie bracket representations of the non-admissible vector fields $f_i^{\neg\mathrm{adm}}$, $i = 1, 2, \ldots, 9$. •

As motivated by the latter example, we aim to determine admissible Lie bracket representations of non-admissible vector fields $f^{\neg\mathrm{adm}} : \mathbb{R}^N \to \mathbb{R}^N$ of the form

$$f^{\neg\mathrm{adm}}(x) = e_j V(\mathbf{x}_i) W([\mathbf{x}_k]_{k \in \mathbb{K}}), \tag{2.50}$$

where $i \in \{1, 2, \ldots, n\}$, $j \in \mathcal{I}(i)$, V, W are smooth real-valued functions and $\mathbb{K}$ is a subset of all agents the ith agent does not have direct access to, i.e.,

$$\mathbb{K} \subseteq \{k \in \{1, 2, \ldots, n\} \mid k \notin \mathcal{N}_{\mathcal{G}}(i)\}. \tag{2.51}$$

As it turns out, given some graph, it mainly depends on $\mathbb{K}$ as well as on the structure of W whether such a non-admissible vector field has an admissible Lie bracket representation. In what follows, we consider three particular cases of increasing difficulty:

Case 1: $\mathbb{K} = \{k\}$ for some $k \in \{1, 2, \ldots, n\}$ and W is an arbitrary real-valued function of $\mathbf{x}_k$;

Case 2: $\mathbb{K}$ is arbitrary and W is *additively separable*, i.e., there exist real-valued functions W_k, $k \in \mathbb{K}$, such that

$$W([\mathbf{x}_k]_{k \in \mathbb{K}}) = \sum_{k \in \mathbb{K}} W_k(\mathbf{x}_k); \tag{2.52}$$

Case 3: $\mathbb{K}$ is arbitrary and W is *multiplicatively separable*, i.e., there exist real-valued functions W_k, $k \in \mathbb{K}$, such that

$$W([\mathbf{x}_k]_{k \in \mathbb{K}}) = \prod_{k \in \mathbb{K}} W_k(\mathbf{x}_k). \tag{2.53}$$

Note that in Example 2 all non-admissible vector fields $f_i^{\neg\mathrm{adm}}$, $i = 1, 2, \ldots, 9$, fall into Case 1. While this is also the most important case for the application at hand, the main motivation to consider Case 2 and Case 3 is to show that the approach also applies in quite general situations by combining the three cases. In the remainder of this section, we first discuss Case 1 to Case 3 separately and then combine these results.

2.3.2.1 Case 1: Vector Fields Depending on the State of a Single Agent

Consider the path $\mathfrak{p}_{1,r}$ as depicted in Figure 2.7. In this section, we aim to find admissible Lie bracket representations of non-admissible vector fields of the form

$$f^{\neg\text{adm}}(x) = e_{j_1} V(\mathbf{x}_1) W(\mathbf{x}_r) \tag{2.54}$$

with $j_1 \in \mathcal{I}(1)$. and where V, W are smooth real-valued functions. To this end, we employ Lemma 1 together with a particular choice of the functions F_{j_k} in (2.43). In particular, comparing (2.54) and (2.42), we need to choose the functions F_{j_k} in (2.43) such that $\tilde{F}_{1,r}$ is a function of $\mathbf{x}_1$ and $\mathbf{x}_r$ only. It is easy to see that this is ensured by choosing

$$F_{j_1}(\mathbf{x}_1, \mathbf{x}_2) = V(\mathbf{x}_1) W_{j_1}(\mathbf{x}_2) \tag{2.55a}$$

$$F_{j_k}(\mathbf{x}_k, \mathbf{x}_{k+1}) = \left(\frac{\partial W_{j_{k-1}}}{\partial x_{j_k}}(\mathbf{x}_k)\right)^{-1} W_{j_k}(\mathbf{x}_{k+1}), \quad k = 2, 3, \ldots, r-1, \tag{2.55b}$$

where $V, \{W_{j_k}\}_{k=1,2,\ldots,r-1}$ are smooth real-valued functions and the functions W_{j_k} fulfill

$$\frac{\partial W_{j_{k-1}}}{\partial x_{j_k}}(\mathbf{x}_{i_k}) \neq 0 \text{ for all } \mathbf{x}_{i_k} \in \mathbb{R}^{|\mathcal{I}(i_k)|}, \quad k = 2, 3, \ldots, r-1. \tag{2.56}$$

Note that less restrictive differentiability assumptions are sufficient for $V, \{W_{j_k}\}_{k=1,2,\ldots,r-1}$; still, we assume the functions to be smooth here for the sake of simplicity. The choice (2.55) then implies that

$$\tilde{F}_{1,r}(x) = V(\mathbf{x}_1) W_{j_{r-1}}(\mathbf{x}_r) \tag{2.57}$$

in (2.43). The next result then follows immediately from Lemma 1; we directly formulate it for paths $\mathfrak{p}_{i_1,i_r}$ instead of $\mathfrak{p}_{1,r}$.

Proposition 1. Consider a graph $\mathcal{G}$ of n nodes $1, 2, \ldots, n$ and let an agent-to-state assignment $\mathcal{I} : \{1, 2, \ldots, n\} \rightrightarrows \{1, 2, \ldots, N\}$ be given. Let $\mathfrak{p}_{i_1,i_r} = \langle i_1 \mid i_2 \mid \ldots \mid i_r \rangle$ denote a simple path in $\mathcal{G}$ from i_1 to i_r. Let a map $\psi : \langle i_k \mid i_{k+1} \rangle \mapsto j_k$, $j_k \in \mathcal{I}(i_k)$, $k = 1, 2, \ldots, \ell(\mathfrak{p}_{i_1,i_r})$, be given. For a given set of smooth real-valued functions $V, \{W_{j_k}\}_{k=1,2,\ldots,r-1}$ fulfilling (2.56), let the functions F_{j_k}, $k = 1, 2, \ldots, \ell(\mathfrak{p}_{i_1,i_r})$, be defined as in (2.55). Then, with $W = W_{j_{r-1}}$,

$$R_\psi(\mathfrak{p}_{i_1,i_r})(x) = e_{j_1} V(\mathbf{x}_{i_1}) W(\mathbf{x}_{i_r}) \tag{2.58}$$

for all $x \in \mathbb{R}^N$ and $R_\psi(\mathfrak{p}_{i_1,i_r})$ is a Lie bracket of admissible vector fields. •

By the latter result, non-admissible vector fields of the form (2.50) falling into Case 1 have an admissible Lie bracket representation for given functions V, W whenever there exists a path from node i to node k. Such admissible Lie bracket representations are highly non-unique; in fact we have the following degrees of freedom:

(A) the choice of the path $\mathfrak{p}_{i,k}$ as well as the subpaths in (2.41),

(B) the choice of the map ψ defined in (2.39), i.e., the choice of the indices j_k, and

(C) the choice of the functions $W_{j_k}, k = 1, 2, \ldots, r-2$, in (2.55).

As to (A), it is preferable to choose the shortest path since the length of the path is equal to the degree of the corresponding admissible Lie bracket representation and, as it turns out later, calculating the approximating input sequences as required in Step 2 of the two-step procedure gets harder with increasing degree. We do not elaborate on how to choose the subpaths at this place but postpone the discussion to Section 2.4, in particular Lemma 5, and consider (B), (C) in the following. As apparent from Proposition 1, given V and W, each specific choice of the set of functions $\{W_{j_k}\}_{k=1,2,\ldots,r-2}$ and the map ψ will result in different admissible vector fields the corresponding Lie bracket is built of. In view of the scheme from Figure 2.3 it is clear that the choice of admissible vector fields does have an effect on the approximating system (2.15); thus, it is desirable to choose them in such a way that, roughly speaking, "the non-distributed algorithm is approximated best". With the approximation procedure being explained only later in Section 2.4, it is understood that it is difficult to catch how approximation quality can be improved at this point. Still, to grasp the gist of the following ideas, it is sufficient to know that the approximating input sequences U_k^σ (cf. Figure 2.3) are of highly oscillatory nature, i.e., they act as perturbations on the distributed algorithm in direction of all required admissible vector fields. Keeping this in mind, we argue that the following two properties are of interest: (i) boundedness of the required admissible vector fields, and (ii) perturbations acting preferably in directions that do not affect the primal variables. As to (i), it is not possible to ensure boundedness of all required admissible vector fields as the following Lemma shows.

Lemma 2. Consider a graph $\mathcal{G}$ of n nodes $1, 2, \ldots, n$ and let an agent-to-state assignment $\mathcal{I} : \{1, 2, \ldots, n\} \rightrightarrows \{1, 2, \ldots, N\}$ be given. Let $\mathfrak{p}_{i_1,i_r} = \langle i_1 \mid i_2 \mid \ldots \mid i_r \rangle$ denote a simple path in $\mathcal{G}$ from i_1 to i_r, $r \geq 3$. Then, for any smooth real-valued functions V, W, with V not identically zero and W non-constant, there exist no map $\psi : \langle i_k \mid i_{k+1} \rangle \mapsto j_k$, $j_k \in \mathcal{I}(i_k)$, $k = 1, 2, \ldots, r-1$, and no corresponding set of smooth bounded functions F_{j_k}, $k = 1, 2, \ldots, r-1$, such that (2.58) holds. •

A proof is provided in Appendix B.1.3. From the proof we additionally infer that at most half of the admissible vector fields in (2.58) can be bounded. A specific choice leading to that property is presented in Michalowsky et al. (2018); we do not elaborate on that since numerical results indicate that this has no benefits but we concentrate on a particularly simple choice also utilized in Ebenbauer et al. (2017); Michalowsky et al. (2017a); Michalowsky, Gharesifard, and Ebenbauer (2020). Therein, the functions chosen are linear and given by

$$W_{j_k}(\mathbf{x}_{i_{k+1}}) = x_{j_{k+1}}, \quad k = 1, 2, \ldots, r-2, \tag{2.59}$$

yielding

$$F_{j_k}(\mathbf{x}_{i_k}, \mathbf{x}_{i_{k+1}}) = \begin{cases} x_{j_2} V(\mathbf{x}_{i_1}) & \text{if } k = 1 \\ x_{j_{k+1}} & \text{if } k = 2, \ldots, r-2 \\ W(\mathbf{x}_{i_r}) & \text{if } k = r-1. \end{cases} \tag{2.60}$$

In the remainder we assume that the functions W_{j_k} are chosen according to (2.60); we emphasize that the subsequent results also apply under different choices.

Remark 3. We mention that the definitions (2.59), (2.60) are not completely precise since, in both equations, the right-hand side depends on k but the left-hand side does not include k as a parameter explicitly. Still, k can be extracted from j_k, i_k or i_{k+1}; we suppose that this is understood to avoid a heavy use of notation. •

As it comes to (ii), in view of the structure of the admissible vector fields (2.38) and keeping in mind that the highly oscillatory perturbations act into the directions of the admissible vector fields according to (2.15), the directions the perturbations are acting in are determined by the indices j_k, i.e., the map ψ. It is an interesting question whether there exists a choice of indices that is in some sense optimal in terms of the approximation quality; we do not address this question in this work but rather propose a heuristic approach and utilize choices that turned out to work well in numerical simulations. For example, since we are most interested in the primal variables, it turned out to be beneficial if the perturbations act on the dual variables if possible. This can be ensured by choosing

$$j_k = \psi(\langle i_k \mid i_{k+1} \rangle) = \max(\mathcal{I}(i_k)), \qquad k = 2, 3, \ldots, r-1, \tag{2.61}$$

where $\max(S)$ denotes the largest value in a set $S \subset \mathbb{N}_{>0}$. In the following we assume that the map ψ is determined according to some not further specified rule (e.g., (2.61)). Together with (2.60), besides the choice of the subpaths, $R_\psi(\mathfrak{p}_{i_1,i_r})$ is then determined by $j_1 = \psi(\langle i_1 \mid i_2 \rangle)$ and the functions $V : \mathbb{R}^{|\mathcal{I}(i_1)|} \to \mathbb{R}$, $W : \mathbb{R}^{|\mathcal{I}(i_r)|} \to \mathbb{R}$. For a path $\mathfrak{p}_{i_1,i_r} = \langle i_1 \mid i_2 \mid \ldots \mid i_r \rangle$ and any $j \in \mathcal{I}(i_1)$, we then define

$$R_j^{V,W}(\mathfrak{p}_{i_1,i_r}) = R_\psi(\mathfrak{p}_{i_1,i_r}), \tag{2.62}$$

where $\psi(\langle i_1 \mid i_2 \rangle) = j$, $\psi(\langle i_k \mid i_{k+1} \rangle) = j_k$, $k = 2, \ldots, r-1$, is given, e.g., by (2.61), and the functions F_{j_k} are defined according to (2.60). This allows us to express (2.58) more conveniently as

$$R_j^{V,W}(\mathfrak{p}_{i_1,i_r})(x) = e_j V(\mathbf{x}_{i_1}) W(\mathbf{x}_{i_r}). \tag{2.63}$$

Before we elaborate on Case 2 and Case 3, we reconsider (2.27) to illustrate how to apply the previous results.

Example 2 (continuing from p. 27). Reconsider the saddle-point dynamics (2.47) and the corresponding non-admissible vector fields (2.49). We utilize the previous results, in particular Proposition 1 and Lemma 3, to determine admissible Lie bracket representations of the non-admissible vector fields. We do not discuss the procedure for all non-admissible vector fields in detail, but exemplarily consider $f_1^{\neg\mathrm{adm}}$ and $f_2^{\neg\mathrm{adm}}$. For $f_1^{\neg\mathrm{adm}}$, we first rewrite it in the form (2.54) as

$$f_1^{\neg\mathrm{adm}}(x) = e_1 x_8 = e_j V(\mathbf{x}_1) W(\mathbf{x}_3), \tag{2.64}$$

where $j = 1 \in \mathcal{I}(1)$, $V(\mathbf{x}_1) = 1$, $W(\mathbf{x}_3) = x_8$. For Proposition 1 to apply, we hence require a path from node 1 to node 3 given by $\mathfrak{p}_{1,3} = \langle 1 \mid 2 \mid 3 \rangle$. Following (2.61), we then let ψ be defined as $\psi(\langle 1 \mid 3 \rangle) = j = 1$, $\psi(\langle 2 \mid 3 \rangle) = 6 \in \mathcal{I}(2)$. By Proposition 1 and using the specific choice (2.60), we infer that

$$f_1^{\neg\mathrm{adm}}(x) = R_1^{V,W}(\mathfrak{p}_{1,3})(x) = \big[e_6 W(\mathbf{x}_3), e_2 x_6 V(\mathbf{x}_1)\big] = [e_6 x_8, e_1 x_6]. \tag{2.65}$$

Note that, with $\mathcal{I}(2) = \{2, 6\}$, we might as well let $\psi(\langle 2 \mid 3 \rangle) = 2$ resulting in the admissible Lie bracket representation $f_1^{\neg\mathrm{adm}}(x) = [e_2 x_8, e_1 x_2]$. As a slightly more complicated example, we next consider the non-admissible vector field $f_2^{\neg\mathrm{adm}}$ which, again, we first rewrite as

$$f_2^{\neg\mathrm{adm}}(x) = e_1 2 x_9 x_1 = e_j V(\mathbf{x}_1) W(\mathbf{x}_4) \tag{2.66}$$

with $j = 1$, $V(\mathbf{x}_1) = 2x_1$, $W(\mathbf{x}_4) = x_9$. Hence, we require a path from node 1 to node 4 which is given by $\mathfrak{p}_{1,4} = \langle 1 \mid 2 \mid 3 \mid 4 \rangle$. We again utilize the specific choice (2.60), (2.61) and obtain

$$f_2^{\neg\mathrm{adm}}(x) = R_1^{V,W}(\mathfrak{p}_{1,4})(x). \tag{2.67}$$

With $\ell(\mathfrak{p}_{1,4}) > 2$, this time the specific structure of the admissible Lie bracket representation depends on the choice of the subpath in the recursion (2.40), (2.41). In fact, we have two different options resulting in the following two different representations:

$$f_2^{\neg\mathrm{adm}}(x) = \Big[e_8 W(\mathbf{x}_4), \big[e_6 x_8, e_1 x_6 V(\mathbf{x}_1)\big]\Big] = \Big[\big[e_8 W(\mathbf{x}_4), e_6 x_8\big], e_1 x_6 V(\mathbf{x}_1)\Big] \tag{2.68}$$

In the following, we only consider the first representation. For the remaining non-admissible vector fields we do not discuss the rewriting procedure in that detail but summarize the resulting admissible Lie bracket representations in Table 2.1. •

2.3.2.2 Case 2: Additively Separable Functions

We next consider non-admissible vector fields of the form

$$f^{\neg\mathrm{adm}}(x) = e_j \sum_{k \in \mathbb{K}} V_k(\mathbf{x}_i) W_k(\mathbf{x}_k) \tag{2.69}$$

with $i \in \{1, 2, \ldots, n\}$, $j \in \mathcal{I}(i)$, $\mathbb{K}$ as in (2.51), and where V_k, W_k are sufficiently smooth real-valued functions. With each summand in (2.69) being a non-admissible vector field of the form (2.54), given that there exists a path from node i to node k for any $k \in \mathbb{K}$, finding admissible Lie bracket representations is straightforward employing Proposition 1. In other words, an additively separable non-admissible vector field of the form (2.69) has an admissible Lie bracket representation for a graph $\mathcal{G} = (\mathcal{V}, \mathcal{E})$ if $\mathbb{K} \subseteq \mathcal{R}_{\mathcal{G}}(i)$ where $\mathcal{R}_{\mathcal{G}}(i)$ is the set of nodes reachable from node i, i.e.,

$$\mathcal{R}_{\mathcal{G}}(i) = \{j \in \mathcal{V} \mid \text{There exists a directed path from node } i \text{ to node } j \text{ in } \mathcal{G}.\}. \tag{2.70}$$

We emphasize that $i \notin \mathcal{R}_{\mathcal{G}}(i)$ for any node i in $\mathcal{G}$.

Non-Admissible Vector Field	Path	Admissible Lie Bracket Representation
$f_1^{\neg\text{adm}}(x) = e_1 x_8$	$\mathfrak{p}_{1,3} = \langle 1 \mid 2 \mid 3 \rangle$	$f_1^{\neg\text{adm}}(x) = [e_6 x_8, e_1 x_6]$
$f_2^{\neg\text{adm}}(x) = e_1 2 x_9 x_1$	$\mathfrak{p}_{1,4} = \langle 1 \mid 2 \mid 3 \mid 4 \rangle$	$f_2^{\neg\text{adm}}(x) = \big[e_8 x_9, [e_6 x_8, 2 e_1 x_6 x_1]\big]$
$f_3^{\neg\text{adm}}(x) = -e_1 x_9 x_4$	$\mathfrak{p}_{1,4} = \langle 1 \mid 2 \mid 3 \mid 4 \rangle$	$f_3^{\neg\text{adm}}(x) = \big[-e_8 x_9, [e_6 x_8, e_1 x_6]\big]$
$f_4^{\neg\text{adm}}(x) = e_2 2 x_2 x_7$	$\mathfrak{p}_{2,1} = \langle 2 \mid 5 \mid 1 \rangle$	$f_4^{\neg\text{adm}}(x) = [e_5 x_7, 2 e_2 x_5 x_2]$
$f_5^{\neg\text{adm}}(x) = -e_4 x_1 x_9$	$\mathfrak{p}_{4,1} = \langle 4 \mid 5 \mid 1 \rangle$	$f_5^{\neg\text{adm}}(x) = [-e_5 x_1, e_4 x_5 x_9]$
$f_6^{\neg\text{adm}}(x) = -e_5 x_6$	$\mathfrak{p}_{5,2} = \langle 5 \mid 1 \mid 2 \rangle$	$f_6^{\neg\text{adm}}(x) = [-e_7 x_6, e_5 x_7]$
$f_7^{\neg\text{adm}}(x) = e_8 x_8 x_1$	$\mathfrak{p}_{3,1} = \langle 3 \mid 4 \mid 5 \mid 1 \rangle$	$f_7^{\neg\text{adm}}(x) = \big[e_5 x_1, [e_9 x_5, e_8 x_9 x_8]\big]$
$f_8^{\neg\text{adm}}(x) = e_9 x_9 x_1^2$	$\mathfrak{p}_{4,1} = \langle 4 \mid 5 \mid 1 \rangle$	$f_8^{\neg\text{adm}}(x) = [e_5 x_1^2, e_4 x_9]$
$f_9^{\neg\text{adm}}(x) = -e_9 x_9 x_1 x_4$	$\mathfrak{p}_{4,1} = \langle 4 \mid 5 \mid 1 \rangle$	$f_9^{\neg\text{adm}}(x) = [-e_5 x_1, e_4 x_4 x_9]$

Table 2.1. An overview of all admissible Lie bracket representations of the non-admissible vector fields $f_i^{\neg\text{adm}}$, $i = 1, 2, \ldots, 9$, occurring in Example 2.

2.3.2.3 Case 3: Multiplicatively Separable Functions

In Case 3, we consider non-admissible vector fields of the form

$$f^{\neg\text{adm}}(x) = e_j V(\mathbf{x}_i) \prod_{k \in \mathbb{K}} W_k(\mathbf{x}_k) \tag{2.71}$$

with $i \in \{1, 2, \ldots, n\}$, $j \in \mathcal{I}(i)$, $\mathbb{K}$ as in (2.51) and where V, W_k are sufficiently smooth real-valued functions. Again, similar as in Case 2, each factor in (2.71) is of the form (2.54); however, in contrast to addition, multiplication cannot be handled directly. Therefore, we first derive an intermediate result that allows to represent a product of functions by means of a Lie bracket of the single factors.

Lemma 3. Consider a graph $\mathcal{G}$ of n nodes $1, 2, \ldots, n$ and let an agent-to-state assignment $\mathcal{I} : \{1, 2, \ldots, n\} \rightrightarrows \{1, 2, \ldots, N\}$ be given. Let further a set of smooth real-valued functions $V, W_k, k \in \mathbb{K} = \{k_1, k_2, \ldots, k_M\}, M \geq 2$, be given, where $\mathbb{K}$ fulfills (2.51). Fix $i \in \{1, 2, \ldots, n\}$, $j \in \mathcal{I}(i)$. For each $k \in \mathbb{K}$, define a vector field $\varphi_k : \mathbb{R}^N \to \mathbb{R}^N$ as

$$\varphi_k(x) = \begin{cases} e_j W_{k_1}(\mathbf{x}_{k_1}) & \text{if } k = k_1 \\ e_j W_{k_M}(\mathbf{x}_{k_M}) \int V(\mathbf{x}_i) \mathrm{d}x_j & \text{if } k = k_M \\ e_j x_j W_k(\mathbf{x}_k) & \text{otherwise,} \end{cases} \tag{2.72}$$

where $\int V(\mathbf{x}_i) \mathrm{d}x_j$ is any antiderivative of V w.r.t. x_j. Then

$$\Big[\Big[\cdots\big[[\varphi_{k_1}, \varphi_{k_2}], \varphi_{k_3}\big], \ldots\Big], \varphi_{k_M}\Big](x) = e_j V(\mathbf{x}_i) \prod_{k \in \mathbb{K}} W(\mathbf{x}_k) \tag{2.73}$$

for all $x \in \mathbb{R}^N$. •

A proof is given in Appendix B.1.4. In general, the vector fields φ_k defined in (2.72) are non-admissible; however, given there exists a path from node i to node k for any $k \in \mathbb{K}$, these non-admissible vector fields have an admissible Lie bracket representation by Proposition 1. Thus, similarly as for Case 2, a multiplicatively separable vector field of the form (2.71) has an admissible Lie bracket representation if $\mathbb{K} \subseteq \mathcal{R}_{\mathcal{G}}(i)$, where $\mathcal{R}_{\mathcal{G}}(i)$ is defined in (2.70). We briefly illustrate this idea by means of an example.

Example 2 (continuing from p. 31). Reconsider Example 2 and suppose that we want to determine an admissible Lie bracket representation of the non-admissible vector field $f_{10}^{\neg\text{adm}}(x) = 2e_1x_8x_9$. Following Lemma 3, we let

$$\varphi_1(x) = e_1x_8, \quad \varphi_2(x) = 2e_1x_1x_9, \tag{2.74}$$

and observe that, for all $x \in \mathbb{R}^N$, we have

$$[\varphi_1, \varphi_2](x) = 2e_1x_8x_9. \tag{2.75}$$

Still, φ_1, φ_2 are non-admissible; however, as discussed in Example 2, we may utilize Proposition 1 to find admissible Lie bracket representations of φ_1, φ_2. In particular, with $\varphi_1 = f_1^{\neg\text{adm}}$, $\varphi_2 = f_2^{\neg\text{adm}}$, we infer from Table 2.1 that

$$f_{10}^{\neg\text{adm}}(x) = 2e_1x_8x_9 = \Big[[e_6x_8, e_1x_6], \big[e_8x_9, [e_6x_8, 2e_1x_6x_1]\big]\Big]; \tag{2.76}$$

hence we have determined an admissible Lie bracket representation of $f_{10}^{\neg\text{adm}}$. •

2.3.2.4 Combinations of Case 2 and Case 3

It is then straightforward to combine Case 2 and Case 3. In fact, for each $i \in \{1, 2, \ldots, n\}$, $j \in \mathcal{I}(i)$, any non-admissible vector field of the form

$$f^{\neg\text{adm}}(x) = e_j \sum_{k=1}^{M} \Big(V_k(\mathbf{x}_i) \prod_{l \in \mathbb{K}} W_{k,l}(\mathbf{x}_l) \Big), \qquad M \in \mathbb{N}_{>0}, \tag{2.77}$$

with $F_k, W_{k,l}$ being real-valued smooth functions, has an admissible Lie bracket representation if $\mathbb{K} \subseteq \mathcal{R}_{\mathcal{G}}(i)$. For a graph $\mathcal{G}$ we then define the set of admissible functions corresponding to node i, $i = 1, 2, \ldots, n$, as

$$\begin{aligned}\mathcal{F}_{\text{adm},\mathcal{G}}(i) := \Big\{ F(x) = \sum_{k=1}^{M} \Big(V_{k_1}(\mathbf{x}_i) \prod_{l \in \mathcal{R}_{\mathcal{G}}(i)} W_{k,l}(\mathbf{x}_l) \Big) \mid M \in \mathbb{N}_{>0}, \\ V_k, W_{k,l}, k \in \{1, 2, \ldots, M\}, l \in \mathcal{R}_{\mathcal{G}}(i), \text{ smooth real-valued functions} \Big\}.\end{aligned} \tag{2.78}$$

With this definition any non-admissible vector field $f^{\neg\text{adm}} = e_j F$, $F \in \mathcal{F}_{\text{adm},\mathcal{G}}(i)$, $j \in \mathcal{I}(i)$, has an admissible Lie bracket representation. When combined with Step 2 of the proposed approach, simply put, this means that an agent i can implement sums and products of functions of the states of the agents that are reachable from agent i by only utilizing the states of its out-neighboring agents. Hence, for a non-distributed algorithm to be implementable in a distributed fashion, each agent is no longer confined to its own state as well as those of its out-neighbors, but may also utilize the states of the agent that are in its reachable set. We admit that this statement is a bit bold since there are several limitations, e.g., the limited structure of the non-admissible vector fields (2.77). Still, (2.77) captures many important cases. In particular, any multivariate polynomial $F(\mathbf{x}_i, [\mathbf{x}_k]_{k\in\mathcal{R}_\mathcal{G}(i)})$ is in $\mathcal{F}_{\text{adm},\mathcal{G}}(i)$. Consequently, if we allow for M to be infinite, then any analytic function of $\mathbf{x}_i$ and $[\mathbf{x}_k]_{k\in\mathcal{R}_\mathcal{G}(i)\}}$ is in $\mathcal{F}_{\text{adm},\mathcal{G}}(i)$. While this observation may not be of practical relevance, still, it shows that admissible Lie bracket representations exist in quite general situations. We note that, for some function F to be in $\mathcal{F}_{\text{adm},\mathcal{G}}(i)$, it is necessary that F is a function of $[\mathbf{x}_k]_{k\in\mathcal{R}_\mathcal{G}(i)}$ and $\mathbf{x}_i$ only. In other words, F may only depend on the corresponding states of all nodes that are reachable from node i as well as the state associated to the node itself.

In case a non-admissible vector field does not fit into the structure (2.77), there is still an alternative based on state augmentation. Simply put, the idea is that each agent carries an estimate of the states of the other agents that may then be used in an arbitrary manner. More precisely, suppose that agent i wants to implement dynamics of the form

$$\dot{\mathbf{x}}_i = \mathbf{f}_i(\mathbf{x}_i, [\mathbf{x}_k]_{k\in\mathcal{R}_\mathcal{G}(i)}). \tag{2.79}$$

We then augment the state by an estimate of $[\mathbf{x}_k]_{k\in\mathcal{R}_\mathcal{G}(i)}$ denoted by μ_i and define the augmented agent dynamics

$$\dot{\mathbf{x}}_i = \mathbf{f}_i(\mathbf{x}_i, \mu_i) \tag{2.80a}$$
$$\dot{\mu}_i = -\theta\mu_i + [\mathbf{x}_k]_{k\in\mathcal{R}_\mathcal{G}(i)} \tag{2.80b}$$

with $\theta > 0$. Then all non-admissible vector fields in the augmented agent dynamics (2.80b) are linear; hence admissible Lie bracket representations exist according to Proposition 1. Clearly, since we have altered the saddle-point dynamics, a stability analysis needs to be carried out; however, from a practical viewpoint, if θ is sufficiently large, singular perturbation arguments apply, i.e., μ_i is in a quasi steady-state and approximately equal to $[\mathbf{x}_k]_{k\in\mathcal{R}_\mathcal{G}(i)}$. Thus, the trajectories of the altered saddle-point dynamics approximate those of the original ones. We additionally note that this state augmentation might even improve the convergence behavior of the distributed algorithm since the state estimation also has a filtering effect which, similar to what is observed in Section 2.5.2, may have a positive effect.

2.3.3 Application to Optimization by means of Saddle-Point Dynamics

In this section, we consider the application of the general results from the previous subsection to the specific dynamics (2.19). We do not aim for determining an admissible Lie bracket representation of the saddle-point dynamics (2.19) for general optimization problems (2.4) but only do this for the special case of convex optimization problems with separable objective function and affine constraints, see Section 2.3.3.2. Although the latter results also apply to the general case, we think that doing so would not provide much insight and is notationally cumbersome; we refer the reader to Example 2 as an illustration how to apply the procedure to more general optimization problems. For general optimization problems (2.4), we are more interested in studying under which assumptions the latter results apply. We discuss this in the following section.

2.3.3.1 General Convex Optimization Problems

Consider a convex optimization problem of the form (2.4) and let some graph $\mathcal{G}$ of n nodes and an agent-to-state assignment $\mathcal{I}$ fulfilling (2.23) be given. The corresponding saddle-point dynamics (2.19) can then be written as

$$\begin{aligned}\dot{x} &= \sum_{i=1}^{n} e_i \tfrac{\partial H}{\partial x_i}(z) - \sum_{i=1}^{n}\sum_{j=1}^{n_{\mathrm{eq}}} e_i \tfrac{\partial a_j}{\partial x_i}(z) x_{n+j} - \sum_{i=1}^{n}\sum_{j=1}^{n_{\mathrm{ineq}}} e_i \tfrac{\partial c_j}{\partial x_i}(z) x_{n+n_{\mathrm{eq}}+j} \\ &\quad + \sum_{i=1}^{n}\sum_{j\in\mathcal{I}_{\mathrm{eq}}(i)} e_{n+j} a_j(z) + \sum_{i=1}^{n}\sum_{j\in\mathcal{I}_{\mathrm{ineq}}(i)} e_{n+n_{\mathrm{eq}}+j} x_{n+n_{\mathrm{eq}}+j} c_j(z) \\ &= f^{\mathrm{adm}}(x) + f^{\neg\mathrm{adm}}(x) \end{aligned} \tag{2.81}$$

with $z = [x_k]_{k\in\{1,2,\ldots,n\}}$. Our goal is then to utilize the previous results to determine an admissible Lie bracket representation of the non-admissible vector field $f^{\neg\mathrm{adm}}$ in (2.81). From our previous discussions it is clear that whether such an admissible Lie bracket representation exist heavily depends on the communication graph as well as the specific optimization problem, i.e., the objective function H as well as the constraints defined by a and c. More precisely, given an optimization problem (2.4), we can formulate conditions on the communication graph to ensure that an admissible Lie bracket representation of $f^{\neg\mathrm{adm}}$ exists. On the other hand, given a graph, we can formulate restrictions on the optimization problem that ensure that $f^{\neg\mathrm{adm}}$ has an admissible Lie bracket representation. We elaborate on that in the remainder of this section. Following our discussions in Section 2.3.2, in particular Section 2.3.2.4, we infer that $f^{\neg\mathrm{adm}}$ has an admissible Lie bracket representation if, for each $i = 1,2,\ldots,n$, the functions

$$F_1^i(x) = \tfrac{\partial H}{\partial x_i}(z) \tag{2.82a}$$

$$F_2^{i,j}(x) = \tfrac{\partial a_j}{\partial x_i}(z) x_{n+j} \qquad j\in\{1,2,\ldots,n_{\mathrm{eq}}\} \tag{2.82b}$$

$$F_3^{i,j}(x) = \tfrac{\partial c_j}{\partial x_i}(z) x_{n+n_{\mathrm{eq}}+j} \qquad j\in\{1,2,\ldots,n_{\mathrm{ineq}}\} \tag{2.82c}$$

$$F_4^{i,j}(x) = a_j(z) \qquad j\in\mathcal{I}_{\mathrm{eq}}(i) \tag{2.82d}$$

$$F_5^{i,j}(x) = c_j(z) x_{n+n_{\mathrm{eq}}+j} \qquad j\in\mathcal{I}_{\mathrm{ineq}}(i) \tag{2.82e}$$

are in $\mathcal{F}_{\mathrm{adm},\mathcal{G}}(i)$. Let $\mathcal{F}(i)$ denote the set of functions defined by (2.82), $i = 1,2,\ldots,n$. For being able to determine admissible Lie bracket representations utilizing the results presented in the previous part of this section, the graph $\mathcal{G}$ must contain the following set of paths:

$$\mathcal{P} = \{\mathfrak{p}_{i,j} \text{ in } \mathcal{G},\, i,j = 1,2,\ldots,n, i \neq j, \mid \exists F \in \mathcal{F}(i), x \in \mathbb{R}^N : \frac{\partial F}{\partial \mathbf{x}_j}(x) \neq 0\}. \tag{2.83}$$

It is clear that this condition is fulfilled if the graph is strongly connected; we utilize this observation to state a sufficient condition for the existence of admissible Lie bracket representations in the following result.

Lemma 4 (Sufficient Condition for the Existence of Admissible Lie Bracket Representations). Consider an optimization problem of the form (2.4) and suppose that there exist functions $H^{k,l}, a_i^{k,l}, c_i^{k,l} : \mathbb{R} \to \mathbb{R}$ and constants $M_H, M_{a,i}, M_{c,i} \in \mathbb{N}$ such that

$$H(z) = \sum_{k=1}^{M_H} \sum_{l=1}^{n} H^{k,l}(z_l) \tag{2.84a}$$

$$a_i(z) = \sum_{k=1}^{M_{a,i}} \sum_{l=1}^{n} a_i^{k,l}(z_l) \qquad i = 1,2,\ldots,n_{\text{eq}} \tag{2.84b}$$

$$c_i(z) = \sum_{k=1}^{M_{c,i}} \sum_{l=1}^{n} c_i^{k,l}(z_l) \qquad i = 1,2,\ldots,n_{\text{ineq}} \tag{2.84c}$$

for all $z \in \mathbb{R}^p$. Then, for any strongly connected graph $\mathcal{G}$ of n nodes and any agent-to-state assignment $\mathcal{I} : \{1,2,\ldots,n\} \rightrightarrows \{1,2,\ldots,N\}$, the saddle-point dynamics (2.19) have an admissible Lie bracket representation. •

Proof. The proof is straightforward and we only provide a sketch. With (2.84), we first note that the right-hand side of (2.19) is a sum of vector fields of the form (2.77) with $\mathbb{K} = \{1,2,\ldots,n\}$, cf. (2.81). As discussed in Section 2.3.2.4, such vector fields have an admissible Lie bracket representation w.r.t. to $(\mathcal{G},\mathcal{I})$ if $\mathbb{K}$ is a subset of the reachable set of the respective node. With $\mathcal{G}$ begin strongly connected, all nodes in the graph are reachable, thus the result follows. □

In most cases, a strongly connected graph is not required but much less restrictive assumptions are sufficient. In fact, if we fix the agent-to-state assignment, then the requirements on the graph are given by (2.83), while the structural requirements on the optimization problem are as formulated in Lemma 4.

In the remainder of this chapter, since required for Step 2 in the general procedure from Figure 2.3, we assume that $\mathcal{F}(i) \subseteq \mathcal{F}_{\text{adm},\mathcal{G}}(i)$, i.e., the saddle-point dynamics have an admissible Lie bracket representation. We may then equally write (2.81) as

$$\dot{x} = f^{\text{adm}}(x) + \sum_{B \in \mathcal{B}} v_B B(x), \tag{2.85}$$

where f^{adm} is the admissible part of (2.81) and $\mathcal{B} \subseteq \mathcal{LB}r(\Phi^{\text{adm}}(\mathcal{G},\mathcal{I}))$ is a subset of Lie brackets generated from some subset of the admissible vector fields $\Phi^{\text{adm}}(\mathcal{G},\mathcal{I})$ and $v_B \in \mathbb{R}$.

We emphasize that each bracket $B \in \mathcal{B}$ has a representation in terms of the recursion (2.40), (2.41), i.e., $B = R_\psi(\mathfrak{p})$ for some ψ, some path $\mathfrak{p} \in \mathcal{G}$ and appropriately chosen functions F_j, or, in case (2.81) contains vector fields (2.71) involving multiplicatively separable functions, i.e., Lemma 3 has been utilized, B is a Lie bracket of brackets represented by that recursion. In the following we will mainly concentrate on the case that (2.81) does not contain non-admissible vector fields of the form (2.71). We note that, in view of (2.82), this is the case if, for each $i = 1, 2, \ldots, n$, there exists $j \in \mathcal{R}_\mathcal{G}(i)$ such that each function $F \in \mathcal{F}$ fulfills

$$\frac{\partial F}{\partial \mathbf{x}_k}(x) = 0 \quad \text{for all } x \in \mathbb{R}^N \tag{2.86}$$

for any $k \in \{1, 2, \ldots, n\} \setminus \{i, j\}$. In particular, this holds if the objective function H is separable and the inequality constraints defined by c are affine; we discuss this important special case in more detail in Section 2.3.3.2. Under this assumption, all non-admissible vector fields take the form (2.69), and, utilizing (2.62), we may write (2.85) as

$$\dot{x} = f^{\text{adm}}(x) + \sum_{\mathfrak{p} \in \mathcal{P}} \sum_{j \in \mathcal{I}(\text{head}(\mathfrak{p}))} v_{B_j(\mathfrak{p})} B_j(\mathfrak{p})(x), \qquad B_j(\mathfrak{p}) = R_j^{V_j(\mathfrak{p}), W_j(\mathfrak{p})}(\mathfrak{p}), \tag{2.87}$$

where $\mathcal{P}$ is the set of all required paths in $\mathcal{G}$ as defined in (2.83), and $V_j(\mathfrak{p}), W_j(\mathfrak{p})$ are smooth real-valued functions determined by the functions (2.82), $i = 1, 2, \ldots, n$. It is indeed possible to state (2.87) (and also (2.85)) explicitly in terms of the functions defining the optimization problem; however, for general optimization problems this requires a heavy use of notation. Thus, while still being notationally cumbersome, we illustrate this for the particular class of convex optimization problems with separable objective function and affine constraints in the next section.

2.3.3.2 Convex Optimization Problems with Separable Objective Function and Affine Constraints

In the following we consider an important subclass of optimization problems of the form (2.4) where, in addition to Assumption 1 and Assumption 2, the objective function H is additively separable, and the inequality constraint c is affine, i.e.,

$$H(z) = \sum_{i=1}^{p} H_i(z_i), \qquad c(z) = Cz - d, \tag{2.88}$$

where $H_i : \mathbb{R} \to \mathbb{R}$, $i = 1, 2, \ldots, p$, $C \in \mathbb{R}^{n_{\text{ineq}} \times p}$, $d \in \mathbb{R}^{n_{\text{ineq}}}$. The application of the proposed approach to these kind of problems is discussed in detail in Michalowsky, Gharesifard, and Ebenbauer (2020). For the sake of a simpler notation, we further assume that the number of inequality and equality constraint is equal to the number of agents, i.e., $n_{\text{eq}} = n_{\text{ineq}} = n$. We note that this also covers the case of $n_{\text{eq}} \leq n$, $n_{\text{ineq}} \leq n$, simply by introducing additional constraints that do not alter the feasible set of the optimization problem, see Michalowsky, Gharesifard, and Ebenbauer (2020) for details. Under this assumption, it is reasonable to associate one constraint to each agent, and we let $\mathcal{I}_{\mathrm{eq}}(i) = i$, $\mathcal{I}_{\mathrm{ineq}}(i) = i$ for each $i = 1, 2, \ldots, n$. This yields the agent-to-state assignment

$$\mathcal{I}(i) = \{i, i+n, i+2n\} \tag{2.89}$$

for $i = 1, 2, \ldots, n$. With $a(z) = Az - b$, the corresponding saddle-point dynamics are

$$\dot{x} = \begin{bmatrix} \dot{z} \\ \dot{\nu} \\ \dot{\lambda} \end{bmatrix} = \begin{bmatrix} -\nabla H(z) - A^\top \nu - C^\top \lambda \\ Az - b \\ \operatorname{diag}(\lambda)(Cz - d) \end{bmatrix} = f^{\mathrm{adm}}(x) + f^{\neg\mathrm{adm}}(x). \tag{2.90}$$

Therein, the admissible and the non-admissible part are given by

$$f^{\mathrm{adm}}(x) = \begin{bmatrix} -\nabla H(z) - A_{\mathrm{adm},2}\nu - C_{\mathrm{adm},2}\lambda \\ A_{\mathrm{adm},1}z \\ \operatorname{diag}(\lambda)C_{\mathrm{adm},1}z \end{bmatrix}, f^{\neg\mathrm{adm}}(x) = \begin{bmatrix} -A_{\mathrm{rest},2}\nu - C_{\mathrm{rest},2}\lambda \\ A_{\mathrm{rest},1}z \\ \operatorname{diag}(\lambda)C_{\mathrm{rest},1}z \end{bmatrix} \tag{2.91}$$

with

$$A_{\mathrm{adm},1} = \theta(A), \quad A_{\mathrm{adm},2} = \theta(A^\top), \quad A_{\mathrm{rest},1} = A - A_{\mathrm{adm},1}, \quad A_{\mathrm{rest},2} = A^\top - A_{\mathrm{adm},2}, \tag{2.92a}$$

$$C_{\mathrm{adm},1} = \theta(C), \quad C_{\mathrm{adm},2} = \theta(C^\top), \quad C_{\mathrm{rest},1} = C - C_{\mathrm{adm},1}, \quad C_{\mathrm{rest},2} = C^\top - C_{\mathrm{adm},2}, \tag{2.92b}$$

where, for any matrix $Q = [q_{ij}]_{i,j=1,2,\ldots,n}$, we let $\theta(Q)$ be defined as

$$\theta(P) = \sum_{i=1}^{n}\sum_{j=1}^{n} \operatorname{sgn}(|g_{ij}|) q_{ij} e_i e_j^\top, \tag{2.93}$$

where $\operatorname{sgn} : \mathbb{R} \to \{-1, 0, 1\}$ is the sign function, i.e., $\operatorname{sgn}(0) = 0$, $\operatorname{sgn}(x) = 1$ for $x > 0$ and $\operatorname{sgn}(x) = -1$ for $x < 0$. In rough words, $\theta(P)$ "sorts out the admissible parts" of some matrix P. We note that in Michalowsky, Gharesifard, and Ebenbauer (2020) it is assumed that $A_{\mathrm{rest},1} = 0$, $C_{\mathrm{rest},1} = 0$, hence rendering the non-admissible part $f^{\neg\mathrm{adm}}$ linear. Still, also in the more general case considered here, $f^{\neg\mathrm{adm}}$ is a sum of non-admissible vector fields of the form (2.54); for $f^{\neg\mathrm{adm}}$ to have an admissible Lie bracket representation w.r.t. $(\mathcal{G}, \mathcal{I})$, we then require the existence of certain paths in $\mathcal{G}$. More precisely, for any $i \in \{1, 2, \ldots, n\}$, the reachable set of node i must fulfill

$$\begin{aligned} \mathcal{R}_{\mathcal{G}}(i) \supseteq \Big\{ &\{j \mid a_{\mathrm{rest},1,ij} \neq 0\} \cup \{j \mid c_{\mathrm{rest},1,ij} \neq 0\} \\ &\cup \{j \mid a_{\mathrm{rest},2,ij} \neq 0\} \cup \{j \mid c_{\mathrm{rest},2,ij} \neq 0\} \Big\}, \end{aligned} \tag{2.94}$$

where $a_{\mathrm{rest},1,ij}, c_{\mathrm{rest},1,ij}, a_{\mathrm{rest},2,ij}, c_{\mathrm{rest},2,ij}$ denote the (i,j)th element of $A_{\mathrm{rest},1}, C_{\mathrm{rest},1}, A_{\mathrm{rest},2}, C_{\mathrm{rest},2}$, respectively, $i, j = 1, 2, \ldots, n$. In particular, as discussed in Lemma 4, if $\mathcal{G}$ is strongly connected this holds for arbitrary A and C. Suppose now that (2.94) holds, let $R_j^{V,W}$ be defined according to (2.62) and let $V(\mathbf{x}_j) = 1$, $W_j^l(\mathbf{x}_j) = x_{ln+j}$, $l = 0,1,2$, $\tilde{V}_j(\mathbf{x}_j) = x_{2n+j}$. We then infer that

$$R_l^{V,W_j^0}(\mathfrak{p}_{i,j})(x) = e_l x_j = e_l z_j \qquad R_l^{V,W_j^1}(\mathfrak{p}_{i,j})(x) = e_l x_{n+j} = e_l v_j \tag{2.95a}$$

$$R_l^{V,W_j^2}(\mathfrak{p}_{i,j})(x) = e_l x_{2n+j} = e_l \lambda_j \qquad R_l^{\tilde{V}_k,W_j^0}(\mathfrak{p}_{i,j})(x) = e_l x_{2n+k} x_j = e_l \lambda_k z_j; \tag{2.95b}$$

hence

$$\begin{aligned} \dot{x} = f^{\mathrm{adm}}(x) &- \sum_{i=1}^{n}\sum_{j=1}^{n} a_{\mathrm{rest},2,ij} R_i^{V,W_j^1}(\mathfrak{p}_{i,j})(x) - \sum_{i=1}^{n}\sum_{j=1}^{n} c_{\mathrm{rest},2,ij} R_i^{V,W_j^2}(\mathfrak{p}_{i,j})(x) \\ &+ \sum_{i=1}^{n}\sum_{j=1}^{n} a_{\mathrm{rest},1,ij} R_{n+i}^{V,W_j^0}(\mathfrak{p}_{i,j})(x) + \sum_{i=1}^{n}\sum_{j=1}^{n} c_{\mathrm{rest},2,ij} R_{n+n_{\mathrm{eq}}+i}^{V,W_j^0}(\mathfrak{p}_{i,j})(x). \end{aligned} \tag{2.96}$$

With that, we conclude our discussions on Step 1 of the general procedure from Figure 2.3 and advance to Step 2 in the following section.

2.4 Construction of Distributed Algorithms

Our main objective in this section is to elaborate on Step 2 of the general procedure (see Figure 2.3) of constructing suitable input functions U_k^σ in (2.15) such that the trajectories of (2.15) uniformly converge to those of the non-distributed dynamics as we increase σ; we make this more precise shortly. To this end, in the remainder of this section we suppose that we have successfully completed Step 1, i.e., we have managed to write the non-distributed dynamics in the form

$$\dot{x} = f^{\mathrm{adm}}(x) + \sum_{B\in\mathcal{B}} v_B B(x), \tag{2.97}$$

where $v_B \in \mathbb{R}$ and $\mathcal{B}$ is a set of Lie brackets built from the set of admissible vector fields $\Phi^{\mathrm{adm}}(\mathcal{G},\mathcal{I}) = \{\phi_1, \phi_2, \ldots, \phi_M\}$.

The following procedure is based on the results presented in Liu (1997a), Sussmann and Liu (1991), Liu (1997b). In Liu (1997b), the relation between the trajectories of a system of the form

$$\dot{x}^\sigma = f_0(x^\sigma) + \sum_{k=1}^{M} \phi_k(x^\sigma) U_k^\sigma(t), \quad x^\sigma(0) = x_0, \tag{2.98}$$

where $f_0, \phi_k : \mathbb{R}^N \to \mathbb{R}^N$, $U_k^\sigma : \mathbb{R} \to \mathbb{R}$, $x_0 \in \mathbb{R}^N$, and the trajectories of an associated *extended system*

$$\dot{x} = f_0(x) + \sum_{B\in\mathcal{B}} v_B B(x), \quad x(0) = x_0, \tag{2.99}$$

is studied, where $\mathcal{B}$ is a finite set of Lie brackets of the vector fields ϕ_k, $k = 1, \ldots, M$, and $v_B \in \mathbb{R}$, $B \in \mathcal{B}$, is the corresponding coefficient. In our setup, the distributed approximation to be designed will play the role of (2.98) with ϕ_k being the admissible vector fields and (2.97) plays the role of (2.99). It is shown in Liu (1997b) that, under a suitable choice of the family of functions U_k^σ, in the following called *approximating input sequences*, the trajectories of (2.98) uniformly converge to those of (2.99) on compact time intervals for increasing σ, i.e., for each $x_0 \in \mathbb{R}^N$, for each $\varepsilon > 0$ and for each $T \geq 0$, there exists $\sigma^* > 0$ such that for all $\sigma > \sigma^*$ and $t \in [0, T]$ we have that

$$\|x(t) - x^\sigma(t)\| \leq \varepsilon. \tag{2.100}$$

A procedure for constructing suitable approximating input sequences U_k^σ that fulfill these assumptions is presented in Liu (1997a) as well as in a brief version in Sussmann and Liu (1991); we will follow this idea in here, however, given that in Liu (1997a) the approximating input sequences are not given in explicit form, we exploit the special structure of the admissible vector fields in order to simplify this procedure and arrive at explicit formulas for a large class of scenarios applicable to our work.

2.4.1 Writing the Lie Brackets in Terms of a P. Hall Basis

The procedure presented in Liu (1997a) requires the brackets used in (2.99) to be brackets in a so-called P. Hall basis; we need to "project" the brackets in (2.97) to such a basis, in the sense that will be made precise shortly. We first recall the definition of a P. Hall basis; we let $\delta(B)$ denote the degree of a bracket B, cf. Appendix A.3.

Definition 5 (P. Hall Basis of a Lie Algebra). Let $\Phi = \{\phi_1, \phi_2, \ldots, \phi_M\}$ be a set of smooth vector fields. A P. Hall basis $\mathcal{PH}(\Phi) = (\mathbb{P}, \prec)$ of the Lie algebra generated by Φ is a set $\mathbb{P}$ of brackets equipped with a total ordering $\prec$ that fulfills the following properties:

[PH1] Every ϕ_k, $k = 1, 2, \ldots, M$, is in $\mathbb{P}$.

[PH2] $\phi_k \prec \phi_j$ if and only if $k < j$.

[PH3] If $B_1, B_2 \in \mathbb{P}$ and $\delta(B_1) < \delta(B_2)$, then $B_1 \prec B_2$.

[PH4] Each $B = [B_1, B_2] \in \mathbb{P}$ if and only if

[PH4.a] $B_1, B_2 \in \mathbb{P}$ and $B_1 \prec B_2$

[PH4.b] either $\delta(B_2) = 1$ or $B_2 = [B_3, B_4]$ for some B_3, B_4 such that $B_3 \preceq B_1$. •

Remark 4. It is understood that a P. Hall basis is well-defined only for formal brackets of indeterminates but not for Lie brackets of vector fields. In particular, in [PH3] and [PH4], for Lie brackets the degree as well as the left and right factors B_1 and B_2 are not uniquely defined, see also Appendix A.3. For the purpose of a clearer presentation we avoid this formal overhead accepting this abuse of notation and assume that B is interpreted as a formal bracket in [PH3], [PH4]. The interested reader is referred to Appendix A.3 for some more details on this subject. •

Note that [PH2] is usually not included in the definition of a P. Hall basis, but it is common to include it for the approximation problem at hand. Moreover, the construction rule [PH4] ensures that no brackets are included in the basis that are related to other brackets in the basis by the Jacobi identity or skew-symmetry; thus the brackets are in this sense independent. However, this does not mean that, when evaluating the brackets, the resulting vector fields are independent, which we will exploit later. It is as well worth mentioning that the ordering fulfilling the properties [PH1]–[PH4] is in general not unique, i.e., for a given set of vector fields Φ, there may exist several P. Hall bases.

Let us now return to our setup. As defined in (2.33), let $\Phi_{\text{all}}^{\text{adm}}(\mathcal{G},\mathcal{I})$ denote the set of all admissible vector fields for a given graph $\mathcal{G}$ and corresponding agent-to-state assignment $\mathcal{I}$. Further, remember that $\Phi^{\text{adm}}(\mathcal{G},\mathcal{I})$ denotes the set of admissible vector fields required for the admissible Lie bracket representations. In view of a shorter notation, we omit the arguments in the following. Every bracket in $\mathcal{LBr}(\Phi^{\text{adm}})$ can then be projected onto some P. Hall basis $\mathcal{PH}(\Phi^{\text{adm}})$, i.e., be uniquely written as a linear combination of elements of $\mathcal{PH}(\Phi^{\text{adm}})$ by successively resorting the brackets, making use of skew-symmetry and the Jacobi identity (cf. (A.6), (A.7)). An illustration of that procedure by means of an example is given in Remark 5. Such a projection algorithm is for example given in Reutenauer (2003) and in the following we let

$$\text{proj}_{\mathbb{P}}(B) = \sum_{\tilde{B}\in\mathbb{P}} \alpha_{\tilde{B}}\tilde{B} \tag{2.101}$$

with $\alpha_{\tilde{B}} \in \mathbb{R}$ denote the unique representation of $B \in \mathcal{LBr}(\Phi^{\text{adm}})$ in terms of brackets from some P. Hall basis $\mathcal{PH}(\Phi^{\text{adm}}) = (\mathbb{P},\prec)$. However, for brackets of higher degree, finding this representation might be tedious and results in a large number of brackets $\tilde{B}$; we hence propose an alternative approach. Instead of resorting the complete brackets appearing in (2.85), we suggest reducing the resorting to brackets of low degree by a proper choice of the subpaths in the construction procedure. More precisely, the main idea is to choose the subpath $\mathfrak{q}$ in (2.41) in such a way that, in each recursion step, the degree of the left factor of the bracket is strictly smaller than the degree of the right factor, hence [PH4.a] holds, and such that the degree of the left factor of the right factor is smaller than that of the left factor of the original bracket, thus [PH4.b] is fulfilled. Since the degree directly corresponds to the length of the subpath this can be achieved by choosing the subpath appropriately, see also Figure 2.8 for an illustration. We make this idea more precise in the following Lemma.

Lemma 5. Consider a graph $\mathcal{G}$ of n nodes denoted by $1,2,\ldots,n$ together with an agent-to-state assignment $\mathcal{I}:\{1,2,\ldots,n\} \rightrightarrows \{1,2,\ldots,N\}$. Let $\Phi^{\text{adm}}(\mathcal{G},\mathcal{I})$ denote the set of required admissible vector fields in (2.97), let some P. Hall basis $\mathcal{PH}(\Phi^{\text{adm}}(\mathcal{G},\mathcal{I})) = (\mathbb{P},\prec)$ be given and let $\text{proj}_{\mathbb{P}}(B)$ denote the unique representation of B in terms of brackets in $\mathbb{P}$, cf. (2.101). For any path $\mathfrak{p}_{i_1,i_r} = \langle i_1 \mid i_2 \mid \ldots \mid i_r\rangle$ in $\mathcal{G}$ and corresponding map ψ as defined in (2.39) we define

$$\tilde{R}_\psi(\mathfrak{p}_{i_1,i_r}) = \begin{cases} R_\psi(\mathfrak{p}_{i_1,i_r}) & \text{if } \ell(\mathfrak{p}_{i_1,i_r}) = 1 \\ \text{proj}_{\mathbb{P}}([R_\psi(\mathfrak{q}^c),R_\psi(\mathfrak{q})]) & \text{if } \ell(\mathfrak{p}_{i_1,i_r}) \in \{2,3,4,6\} \\ [\tilde{R}_\psi(\mathfrak{q}^c),\tilde{R}_\psi(\mathfrak{q})] & \text{otherwise,} \end{cases} \tag{2.102}$$

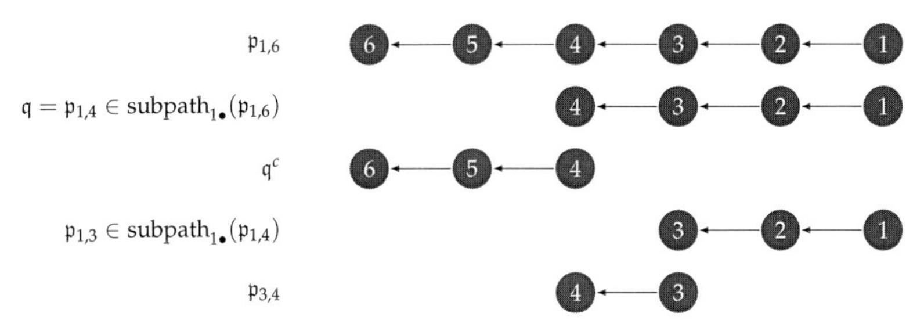

Figure 2.8. An illustration of the idea of choosing the subpaths. The complement of the subpath $\mathfrak{q}^c$ is strictly shorter than the subpath $\mathfrak{q}$ such that in the recursion (2.41) the left factor of the bracket has strictly smaller degree than the right factor, hence [PH4.a] in the Definition of a P. Hall basis holds. Also, the subpath $\mathfrak{p}_{3,4}$ of the subpath $\mathfrak{q}$ is strictly shorter than $\mathfrak{q}^c$ such that in the recursion (2.41) the left factor of the right factor of the bracket has strictly smaller degree than the left factor of the bracket, thus making sure that [PH4.b] holds as well.

where

$$\mathfrak{q} = \mathfrak{p}_{i_1, i_{\theta(p_{i_1 i_r})}} \in \text{subpath}_{i_1\bullet}(\mathfrak{p}_{i_1,i_r}) \tag{2.103a}$$

$$\theta(\mathfrak{p}_{i_1,i_r}) = \begin{cases} \frac{1}{2}\ell(\mathfrak{p}_{i_1,i_r}) + 1 & \text{if } \ell(\mathfrak{p}_{i_1,i_r}) \in \{2,4\} \\ \lfloor \frac{1}{2}\ell(\mathfrak{p}_{i_1,i_r}) \rfloor + 2 & \text{otherwise,} \end{cases} \tag{2.103b}$$

with $\lfloor a \rfloor$ being the largest integer value less or equal than $a \in \mathbb{R}_{\geq 0}$. Then

$$\tilde{R}_\psi(\mathfrak{p}_{i_1,i_r})(x) = R_\psi(\mathfrak{p}_{i_1,i_r})(x) \tag{2.104}$$

for all $x \in \mathbb{R}^N$ and $\tilde{R}_\psi(\mathfrak{p}_{i_1,i_r}) \in \mathbb{P}$. •

A proof is given in Appendix B.1.5. Equation (2.102) and the choice of $\mathfrak{q}$ from (2.103a) can be interpreted as follows: A bracket corresponding to a path $\mathfrak{p}_{i_1,i_r}$ of length larger than one is generated by dividing the path into two complementing subpaths $\mathfrak{q}$ and $\mathfrak{q}^c$, where (2.103b) ensures that the resulting brackets have the desired properties [PH4]. The cases where these properties are not ensured by that choice, i.e., $\ell(\mathfrak{p}_{i_1,i_r}) \in \{2,3,4,6\}$, are handled separately utilizing some projection procedure. It should be noted that in this case the projection can be computed easily; we explain this in the following remark.

Remark 5. Consider (2.40), (2.41) and suppose that the subpaths are chosen according to (2.103). For $\ell(\mathfrak{p}_{i_1,i_r}) = 2, 3$, the brackets admit the following structure

$$R_\psi(\mathfrak{p}_{i_1,i_r}) = \begin{cases} [\phi_{a_1}, \phi_{a_2}] & \text{if } \ell(\mathfrak{p}_{i_1,i_r}) = 2 \\ [\phi_{a_1}, [\phi_{a_2}, \phi_{a_3}]] & \text{if } \ell(\mathfrak{p}_{i_1,i_r}) = 3 \end{cases} \tag{2.105}$$

for some $a_{1/2/3} \in \mathbb{N}_{>0}$ depending on $\psi, \mathfrak{p}_{i_1,i_r}$, where $\phi_{a_i} \in \Phi$, $i = 1,2,3$. For such brackets, the projection on the P. Hall basis $\mathcal{PH}(\Phi) = (\mathbb{P}, \prec)$ is easily computed making use of skew-symmetry (A.6) and the Jacobi-identity (A.7). In fact, we have

$$\mathrm{proj}_{\mathbb{P}}([\phi_{a_1}, \phi_{a_2}]) = \begin{cases} [\phi_{a_1}, \phi_{a_2}] & \text{if } a_1 < a_2, \\ -[\phi_{a_2}, \phi_{a_1}] & \text{if } a_1 > a_2, \end{cases} \tag{2.106}$$

and

$$\mathrm{proj}_{\mathbb{P}}\big([\phi_{a_1}, [\phi_{a_2}, \phi_{a_3}]]\big) = \begin{cases} -\big[\phi_{a_3}, [\phi_{a_1}, \phi_{a_2}]\big] & \text{if } a_1 = \min\limits_{i=1,2,3} a_i, \\ \big[\phi_{a_1}, [\phi_{a_2}, \phi_{a_3}]\big] & \text{if } a_2 = \min\limits_{i=1,2,3} a_i, \\ -\big[\phi_{a_1}, [\phi_{a_3}, \phi_{a_2}]\big] & \text{if } a_3 = \min\limits_{i=1,2,3} a_i, \end{cases} \tag{2.107}$$

where we utilized the Jacobi-identity in the case $a_1 = \min_{i=1,2,3} a_i$ as well as the fact that $\big[\phi_{a_2}, [\phi_{a_1}, \phi_{a_3}]\big] = 0$ as discussed in Remark 2. Note that the brackets have been resorted in such a way that the brackets on the right-hand side of (2.106), (2.107) fulfill [PH3], [PH4] when interpreted as formal brackets. In the same manner, for $\ell(\mathfrak{p}_{i_1,i_r}) = 4, 6$, we have

$$R_\psi(\mathfrak{p}_{i_1,i_r}) = \begin{cases} [B_{a_1}, B_{a_2}] & \text{if } \ell(\mathfrak{p}_{i_1,i_r}) = 4, \\ \big[B_{a_1}, [B_{a_2}, B_{a_3}]\big] & \text{if } \ell(\mathfrak{p}_{i_1,i_r}) = 6, \end{cases} \tag{2.108}$$

where the B_{a_i} are Lie brackets of the ϕ_i with $\delta(B_{a_i}) = 2$, $i = 1,2,3$. The projection is then done by first projecting the inner brackets B_{a_i} on the P. Hall basis using (2.106) and then resorting $R_\psi(\mathfrak{p}_{i_1,i_r})$ as in (2.106), (2.107). •

We note that for non-admissible vector fields in the form (2.71) involving multiplicatively separable functions, we may apply Lemma 5 for each non-admissible vector field ψ_{k_i} in (2.73); still, we need to rewrite the complete resulting bracket in terms of a P. Hall basis using some algorithm, e.g., as given in Reutenauer (2003). Summarizing, utilizing Lemma 5 as well as an appropriate projection procedure (2.101), we may equivalently write (2.97) as

$$\dot{x} = f^{\mathrm{adm}}(x) + \sum_{B \in \mathbb{P}} \tilde{v}_B B(x), \tag{2.109}$$

where each bracket B is an element of a P. Hall basis $\mathcal{PH}(\Phi^{\mathrm{adm}}(\mathcal{G},\mathcal{I})) = (\mathbb{P}, \prec)$ and $\tilde{v}_B \in \mathbb{R}$ is the corresponding coefficient determined from the set of coefficients $\{v_B\}_{B \in \mathcal{LBr}(\Phi^{\mathrm{adm}}(\mathcal{G},\mathcal{I}))}$. Further, if (2.86) holds, i.e., the saddle-point dynamics do not contain non-admissible vector fields (2.71) involving multiplicatively separable functions, then (2.87) can equally be written as

$$\dot{x} = f^{\mathrm{adm}}(x) + \sum_{\mathfrak{p} \in \mathcal{P}} \sum_{j \in \mathcal{I}(\mathrm{head}(\mathfrak{p}))} \tilde{v}_{B_j(\mathfrak{p})} B_j(\mathfrak{p})(x), \qquad B_j(\mathfrak{p}) = \tilde{R}_j^{V_j(\mathfrak{p}),W_j(\mathfrak{p})}(\mathfrak{p}), \tag{2.110}$$

where $\tilde{R}_j^{V,W}(\mathfrak{p})$ is defined as in (2.62) but replacing R_ψ by $\tilde{R}_\psi$. We are now ready to apply the procedure presented in Liu (1997a) to construct suitable approximating input sequences; this will be the main subject of the following section.

2.4.2 Approximating Input Sequences

Let the distributed approximation of (2.97) to be designed be given by

$$\dot{x}^\sigma = f^{\text{adm}}(x^\sigma) + \sum_{k=1}^{M} \phi_k(x^\sigma) U_k^\sigma(t). \tag{2.111}$$

The primary subject of this section is to construct the approximating input sequences U_k^σ in such a way that the solutions of (2.111) uniformly converge to those of (2.109) with increasing σ. To this end, we employ the computation procedure from Liu (1997a). This procedure relies on a "superposition principle", which we briefly sketch in the following. The idea is to group all brackets B in (2.109) into equivalence classes (we define the equivalence relation later), treat each equivalence class separately to obtain a corresponding approximating input, and sum up the resulting approximating inputs in the end. More precisely, we associate to each equivalence class E an input $U_{k,E}^\sigma$ and then let

$$U_k^\sigma(t) = \sum_{E \in \mathcal{E}} U_{k,E}^\sigma(t), \tag{2.112}$$

where $\mathcal{E}$ is the set of all equivalence classes in the set of brackets $\mathbb{P}$. Roughly speaking, two brackets are said to be equivalent if each vector field appears the same number of times in the bracket but possibly in a different order. A precise definition of the equivalence relation is given in Definition 17. For each equivalence class $E \in \mathcal{E}$ and $k = 1, \ldots, M$, we then define the corresponding input $U_{k,E}^\sigma$ as follows:

$$U_{k,E}^\sigma(t) = \begin{cases} 0 & \text{if } \delta_k(E) = 0 \\ 2\sqrt{\sigma}\mathrm{Re}\big(\eta_{E,k}(\omega_E) e^{\mathrm{i}\sigma\omega_E t}\big) & \text{if } \delta_k(E) = 1, \delta(E) = 2 \\ 2\sigma^{N-1/N} \sum_{\rho=1}^{|E|} \mathrm{Re}\big(\eta_E(\omega_{E,\rho,k}) e^{\mathrm{i}\sigma\omega t}\big) & \text{if } \delta_k(E) = 1, \delta(E) = N, N \geq 3, \end{cases} \tag{2.113}$$

where $\mathrm{i} \in \mathbb{C}$ is the imaginary unit and $\mathrm{Re}(x)$ denotes the real part of a complex number $x \in \mathbb{C}$. Here, we let $\delta(E) = \delta(B)$, $\delta_k(E) = \delta_k(B)$ for any $B \in \mathcal{E}$. Further, $\omega_E, \omega_{E,\rho,k} \in \mathbb{R}$ are frequencies we will specify later, $\eta_{E,k}, \eta_E : \mathbb{R} \to \mathbb{C}$ are coefficients to be chosen in dependence of the frequencies, and $\mathrm{i} \in \mathbb{C}$ is the imaginary unit. However, the superposition principle does not hold as desired and there are two major issues one has to take care of:

1. The input sequences $U_{k,E}^\sigma$ may not interfere with each other in order to ensure that the superposition principle holds; this can be dealt with by a proper choice of the frequencies.

2. Each input sequence $U_{k,E}^\sigma$ not only generates the desired brackets $E \cap \mathcal{B}$ for $\sigma \to \infty$, but also all other equivalent brackets in E; we can overcome this by a proper choice of the coefficients $\eta_{\omega,k}, \eta_\omega$. The idea behind this is to also generate the undesired equivalent brackets on purpose – which itself also generate the desired brackets – in such a way that the undesired equivalent brackets all cancel out.

While the problem at hand does not allow for simplifications in the choice of the frequencies, the calculation of proper coefficients $\eta_{\omega,k}, \eta_\omega$ can be simplified drastically by exploiting some structural properties of the set of brackets $\mathcal{B}$. More precisely, there are two properties that turn out to be beneficial: First, in each bracket $B \in \mathcal{B}$ each vector field ϕ_k appears only once, i.e., $\delta_k(B) \in \{0,1\}$, for any $B \in \mathcal{B}, k = 1,\ldots,M$, and second, for any bracket $B \in \mathcal{B}$, all equivalent brackets either evaluate to the same vector field as B or vanish, see Figure C.1 in Appendix C.1.1 as well as Michalowsky, Gharesifard, and Ebenbauer (2017b, Lemma 4) for a proof. We present and discuss the simplified computation procedure exploiting these observations in Appendix C.1.1. While the calculation of the approximating inputs may be tedious, it is not time-consuming, can be done off-line and is algorithmically implementable (cf. Michalowsky et al. (2017b) for an exemplary implementation in MATLAB). It is worth mentioning that the calculation of the P. Hall basis as well as the approximating input sequences is not distributed and requires preliminary global information; hence, the *design* of the distributed algorithm is not distributed but its implementation is. It is a matter of future research to develop distributed design procedures.

2.4.2.1 Explicit Representation of Approximating Input Sequences for Brackets of Low Degree

While the computation procedure given in Appendix C.1.1 can in general be complicated to implement, the procedure becomes particularly simple in scenarios where the set of brackets $\mathbb{P}$ in (2.109) only contains brackets of degree less or equal than three. In a nutshell, in this case the set of equivalent brackets that do not evaluate to zero is a singleton having only the bracket itself as an element, thus the second issue in Section 2.4.2 does not come into play. We prove this in the next result for the simplified representation (2.110). We emphasize that a similar result can be derived for (2.109) with little more effort; we do not do this here since this requires a heavy use of notation.

Proposition 2. Consider a graph $\mathcal{G}$ with nodes $1,2,\ldots,n$ and let an agent-to-state assignment $\mathcal{I} : \{1,2,\ldots,n\} \rightrightarrows \{1,2,\ldots,N\}$ be given. Let $\mathcal{P}$ denote the set of required paths as defined in (2.83) and suppose that $\ell(\mathfrak{p}) \leq 3$ for any path $\mathfrak{p} \in \mathcal{P}$. Let $h_{k,j}$ be defined as in (2.38), let $\Phi^{\mathrm{adm}}(\mathcal{G},\mathcal{I})$ denote the set of required admissible vector fields and let $\mathcal{PH}(\Phi^{\mathrm{adm}}(\mathcal{G},\mathcal{I})) = (\mathbb{P},\prec)$ be any P. Hall basis of $\Phi^{\mathrm{adm}}(\mathcal{G},\mathcal{I})$ that fulfills

$$h_{i_1,\psi(\mathfrak{p}_{i_1,i_2})} \prec h_{i_2,\psi(\mathfrak{p}_{i_2,i_3})} \qquad \text{or} \qquad h_{i_3,\psi(\mathfrak{p}_{i_3,i_4})} \prec h_{i_2,\psi(\mathfrak{p}_{i_2,i_3})} \tag{2.114}$$

for any path $\mathfrak{p}_{i_1,i_4} = \langle i_1 \mid i_2 \mid i_3 \mid i_4 \rangle$ of length three in $\mathcal{P}$. Then, for any $\mathfrak{p} \in \mathcal{P}$ and any $j \in \mathcal{I}(\mathrm{head}(\mathfrak{p}))$, the equivalence class corresponding to the bracket $B_j(\mathfrak{p})$ defined in (2.110) fulfills

$$E_{B_j(\mathfrak{p})} = \{B \in \mathbb{P} : B \sim B_j(\mathfrak{p}), B(z) \not\equiv 0\} = \{B_j(\mathfrak{p})\}, \tag{2.115}$$

where the equivalence relation $\sim$ is defined by Definition 17. •

A proof of this result can be found in Appendix B.1.6. It should be noted that the ordering of the P. Hall basis according to (2.114) is important for this result to hold. A sufficient condition for the requirement that all paths $\mathfrak{p}$ in $\mathcal{P}$ are of length less or equal than three to hold is, for example, that the longest cordless cycle in $\mathcal{G}$ is of length 4. Using the result of Proposition 2 and following the procedure presented in Appendix C.1.1, we obtain the following explicit formulas for the approximating input sequences.

- For $E = \{B\} = \{[\phi_{k_1}, \phi_{k_2}]\}$, we let

$$U^{\sigma}_{k,E}(t) = \begin{cases} -\sqrt{2\sigma}\frac{1}{\beta_E}\sqrt{|v_B\omega_E|}\cos(\sigma\omega_E t) & \text{if } k = k_1 \\ \operatorname{sgn}(v_B\omega_E)\sqrt{2\sigma}\beta_E\sqrt{|v_B\omega_E|}\sin(\sigma\omega_E t) & \text{if } k = k_2 \\ 0 & \text{otherwise.} \end{cases} \tag{2.116}$$

- For $E = \{B\} = \{[\phi_{k_1}, [\phi_{k_2}, \phi_{k_3}]]\}$, we let

$$U^{\sigma}_{k,E}(t) = \begin{cases} \sigma^{\frac{2}{3}}2\beta_E|\frac{1}{2}v_B\omega_{E,k_1}\omega_{E,k_2}|^{\frac{1}{3}}\cos(\sigma\omega_{E,k}t) & \text{if } k = k_1, k_3 \\ \operatorname{sgn}(-v_B\omega_{E,k_1}\omega_{E,k_2})\sigma^{\frac{2}{3}}2\frac{1}{\beta_E^2}|\frac{1}{2}v_B\omega_{E,k_1}\omega_{E,k_2}|^{\frac{1}{3}}\cos(\sigma\omega_{E,k}t) & \text{if } k = k_2 \\ 0 & \text{otherwise.} \end{cases} \tag{2.117}$$

Therein, $\beta_E \neq 0$ is a design parameter; since each approximating input sequence U^{σ}_k is associated to an admissible vector field ϕ_k, this parameter can be used to "distribute the perturbations among the admissible vector fields". According to the copmutation procedure from Appendix C.1.1, the frequencies $\omega_E, \omega_{E,k} \in \mathbb{R} \setminus \{0\}$ need to be chosen such that they fulfill the following properties:

- All frequencies ω_E, $E \in \mathcal{E}$, $\delta(E) = 2$, are distinct.

- For each $E = \{B\} = [\phi_{k_1}, [\phi_{k_2}, \phi_{k_3}]]$, the set of frequencies $\{\omega_{E,k_1}, \omega_{E,k_2}, \omega_{E,k_3}\}$ is minimally canceling, see Definition 18.

- The collection of sets

$$\{\{\omega_E\}_{E\in\mathcal{E},\delta(E)=2}, \{\omega_{E,k_1}, \omega_{E,k_2}, \omega_{E,k_3}\}_{E\in\mathcal{E},\delta(E)=3}\}$$

  is an independent collection, see Definition 19.

Similar explicit formulas can be obtained for brackets of higher degree but they become more complicated. The main reason is that, while for brackets of degree strictly less than four all equivalent brackets evaluate to zero (cf. Figure C.1), this is no longer the case for brackets of higher degree such that now the second issue discussed in Section 2.4.2 needs to be taken care of.

2.4.3 Distributed Algorithm

Employing the approximating input sequences U_k^σ presented in the previous subsection, it is then straightforward to construct a distributed approximation of a non-distributed algorithm that has been written in the form (2.109). In fact, since the control law (2.113) is obtained from the construction procedure presented in Liu (1997a), the following result which sums up our previous considerations follows immediately from Liu (1997a, Theorem 8.1) and Lemma 4.

Theorem 2. Let an optimization problem (2.4) fulfilling Assumption 1, Assumption 2 together with a graph $\mathcal{G}$ of $n = p$ nodes and an agent-to-state assignment $\mathcal{I} : \{1, 2, \dots, n\} \rightrightarrows \{1, 2, \dots, N\}$ defined by (2.23) be given. Consider the saddle-point dynamics (2.19), let $f^{\text{adm}} : \mathbb{R}^N \to \mathbb{R}^N$ denote the admissible part therein and let $\Phi_{\text{all}}^{\text{adm}}(\mathcal{G}, \mathcal{I})$ defined by (2.33) denote the set of admissible vector fields. Suppose that $\mathcal{G}$ is strongly connected and let all conditions from Lemma 4 be fulfilled. Then the following holds.

(i) *(Existence of admissible Lie bracket representation)* There exists a set of admissible vector fields $\{\phi_1, \dots, \phi_M\} \subset \Phi_{\text{all}}^{\text{adm}}(\mathcal{G}, \mathcal{I})$ and a set of Lie brackets $\mathcal{B} \subseteq \mathbb{P}$, where $\mathbb{P}$ is a set of Lie brackets in some P. Hall basis $(\mathbb{P}, \prec)$ of $\{\phi_1, \dots, \phi_M\}$, such that (2.19) can equivalently be written as

$$\dot{x} = f^{\text{adm}}(x) + \sum_{B \in \mathcal{B}} v_B B(x), \tag{2.118}$$

where $v_B \in \mathbb{R}$, $B \in \mathcal{B}$.

(ii) *(Distributed approximation)* For each $\varepsilon > 0$, for each $T > 0$, and for each initial condition $x^\sigma(0) = x(0) = x_0 \in \mathcal{R}(\mathcal{M})$, with $\mathcal{R}(\mathcal{M})$ given in (2.20), there exists $\sigma^* > 0$ such that for all $\sigma > \sigma^*$ the following holds: For all $0 \leq t \leq T$, we have

$$\left\| x^\sigma(t) - x(t) \right\| \leq \varepsilon, \tag{2.119}$$

where $x^\sigma(t)$ is the solution of the distributed algorithm (2.111) with the approximating input sequences (2.113), where the parameters are chosen according to the algorithm presented in Appendix C.1.1, and $x(t)$ is the solution of (2.19), where the initial conditions are given by $x^\sigma(0) = x(0) = x_0$, respectively. •

We note that (i) can be associated to Step 1 and (ii) to Step 2 in the general procedure from Figure 2.3 and the latter Theorem provides a solution to the considered algorithm design problem. We elaborate a bit more on that result. While (i) is immediate from Lemma 4, (ii) is a direct consequence of (i) and Liu (1997a, Theorem 8.1) since the approximating input sequences (2.113) are obtained from the construction procedure presented in the same reference. Concerning (i), we emphasize once more that a strongly connected graph as well as the conditions from Lemma 4 are only sufficient but not necessary. Concerning the distributed approximation (ii), indeed we only require the existence of some admissible Lie bracket representation (2.119) of the saddle-point dynamics (2.19). It is important to mention that (ii) provides convergence of the trajectories on any arbitrary *finite* time interval $[0, T]$. It is fair to say that this is sufficient for the present application since optimization algorithms are typically stopped after some finite time given a desired precision has been achieved. Still, in particular in distributed stabilization problems, one is often interested in providing stability guarantees, too, hence requiring an extension of Theorem 2. To this end, one might be tempted to follow the lines of the proof of Dürr et al. (2013, Theorem 2). Therein, Lie bracket approximations of dynamics involving Lie brackets of degree two are studied. Assuming that a set of equilibria is asymptotically stable for these dynamics and utilizing the trajectory convergence on finite time intervals, it is proven that the set of equilibria is practically asmyptotically stable w.r.t. σ for the approximating dynamics. In the problem at hand, the required asymptotic stability is ensured by Theorem 1 and trajectory converge on finite time intervals by Theorem 2. However, for the line of argumentation from Dürr et al. (2013) to apply, we are required to find *one* σ^* that works for *all* initial conditions $x_0 \in \mathcal{R}(\mathcal{M})$, but the construction procedure from Liu (1997a) is not shown to provide this property. Hence, we cannot follow the lines of the proof of Dürr et al. (2013, Theorem 2). Still, in Dürr et al. (2013) it has been proven that this requirement is fulfilled for approximating input sequences associated to brackets of degree two, and these ideas have been extended to Lie brackets of degree three in Labar, Garone, Kinnaert, and Ebenbauer (2019) utilizing a different construction procedure for the approximating input sequences. We expect that a generalization to brackets of arbitrary degree is possible and conjecture that the construction procedure from Liu (1997a) provides the required uniformity properties. If our conjecture is true, the line of argumentation of the proof of Dürr et al. (2013, Theorem 2) would apply rendering the set of saddle-points $\mathcal{M}$ practically uniformly asymptotically stable (see Dürr et al. (2013) for a definition). Proving this is not within the scope of this thesis; still, in the special case that all admissible Lie brackets are of degree two, it follows from Dürr et al. (2013) that the set $\mathcal{M}$ is is practically uniformly asymptotically stable for (2.111) with the approximating input sequences (2.113).

In the remainder of this section, we finally reconsider Example 2 to illustrate our results. We utilize the construction procedure from Section 2.4 to design a distributed algorithm. We do not explain the construction in detail but refer the reader to the implementation available from Michalowsky et al. (2017b). The numerical results are depicted in Figure 2.9. As expected, the trajectories of the non-distributed saddle-point dynamics converge to the optimal solution of (2) given by

$$z^\star = \begin{bmatrix} 0 & 2 & 2 & 4 & 4 \end{bmatrix}^\top, \nu^\star = -1, \lambda^\star = \begin{bmatrix} 0.5 & 1 & 0 \end{bmatrix}^\top, \tag{2.120}$$

while the trajectories of the corresponding distributed approximation converge to a neighborhood thereof. Notably, the amplitude of the oscillations induced by the periodic perturbations is larger for the dual variables ν, λ compared to the primal variable z. This is mainly caused by the specific choice (2.61) that, as explained beforehand, favors the perturbation of the dual variables. Still, highly oscillatory dual variables might be undesired; as a remedy, we propose adding suitably designed filters. We discuss this extension later in Section 2.5; for a better comparison we depict the numerical results in Figure 2.10.

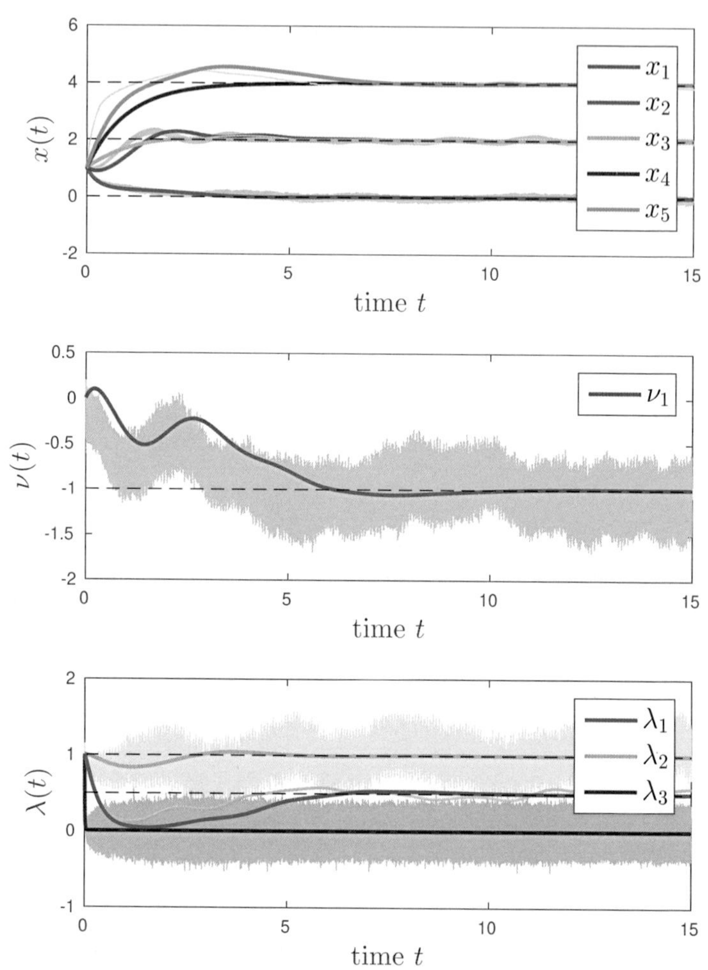

Figure 2.9. Numerical results for Example 2 with initial condition $z(0) = \mathbf{1}$, $\nu(0) = 0$, $\lambda(0) = \mathbf{1}$. The thick lines depict the solution of the non-distributed saddle-point dynamics (2.27), the oscillating lighter lines depict the solution of the distributed approximation with $\sigma = 2000$.

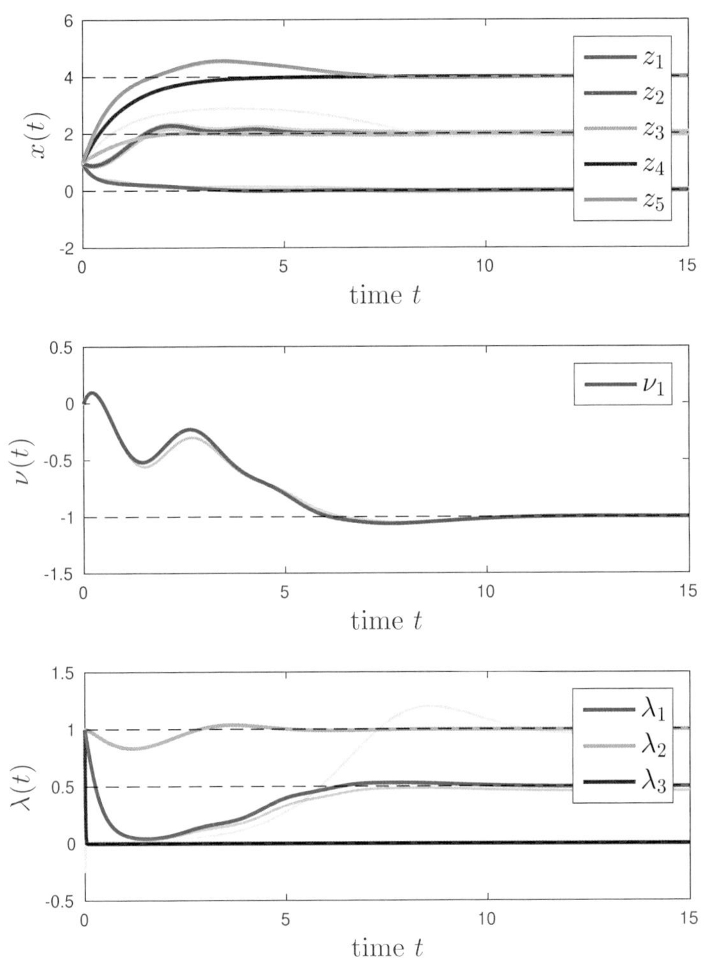

Figure 2.10. Numerical results for Example 2 with initial condition $z(0) = \mathbf{1}$, $\nu(0) = 0$, $\lambda(0) = \mathbf{1}$. The thick lines depict the solution of the non-distributed saddle-point dynamics (2.27), the lighter lines depict the solution of the distributed approximation with $\sigma = 2000$ using first order low-pass filters in order to dampen the oscillations visible in Figure 2.9. The trajectories of the non-distributed saddle-point dynamics and the distributed approximation thereof overlap to a large extent; hence the trajectories of the distributed approximation are partly not visible.

2.5 Discussions

In this section we briefly discuss several possible directions of future research and extensions to the framework presented in this chapter. We emphasize that all what follows should be taken as first ideas.

2.5.1 Choice of Agent-To-State Assignments

Suppose we have given some optimization problem (2.4) as well as a dynamics

$$\dot{x} = f(x) \tag{2.121}$$

with $f : \mathbb{R}^N \to \mathbb{R}^N$ that is able to solve the optimization problem. As already mentioned in Section 2.1, if we want to follow the procedure from Figure 2.3 to derive distributed dynamics from (2.121) given some graph $\mathcal{G}$ or even only check whether (2.121) is distributed according to Definition 1, we need to first assign the components x_i, $i = 1, 2, \ldots, N$, of the state to the agents. In this section we elaborate on this preliminary step for the saddle-point dynamics (2.19). If the number of agents n is equal to the dimension p of the optimization variable as it was assumed in the previous sections, it is reasonable to assign the component z_i to the ith agent. If $n < p$, it is also possible to assign multiple components of z to each agent. We do not consider this here; the procedure is similar to the assignment of the remaining components that we discuss in the following. Since these remaining components are dual variables associated to a constraint in the optimization problem (2.4), the assignment can be interpreted as an assignment of the constraints to the agents. Let $\mathcal{I}_{\mathrm{eq}}(i) \subseteq \{1, 2, \ldots, n_{\mathrm{eq}}\}$, $\mathcal{I}_{\mathrm{ineq}}(i) \subseteq \{1, 2, \ldots, n_{\mathrm{ineq}}\}$ denote the pairwise disjoint index sets of equality and inequality constraints assigned to agent i, i.e., $[a_k]_{k \in \mathcal{I}_{\mathrm{eq}}(i)}$, $[c_k]_{k \in \mathcal{I}_{\mathrm{ineq}}(i)}$ and the corresponding Lagrange multipliers $[\nu_k]_{k \in \mathcal{I}_{\mathrm{eq}}(i)}$, $[\lambda_k]_{k \in \mathcal{I}_{\mathrm{ineq}}(i)}$ are assigned to agent i. It is then reasonable to choose the assignment in such a way that the constraints assigned to agent i are a function of z_i, the component of the primal variable assigned to agent i. This can be ensured by the following assumption.

Assumption 3 (Association of constraints). For each $i \in \{1, 2, \ldots, n\}$ and each $j \in \mathcal{I}_{\mathrm{eq}}(i)$, $k \in \mathcal{I}_{\mathrm{ineq}}(i)$, there exists $z \in \mathbb{R}^n$ such that $\frac{\partial}{\partial z_i} a_j(z) \neq 0$ and $\frac{\partial}{\partial z_i} c_k(z) \neq 0$. •

For implementation purpose, it often makes sense to automate the process of assigning the constraints to the agents. Since each information not directly available to an agent from its neighbors needs to be approximately generated, it is reasonable to choose the assignment in such a way that the effort of generating this missing information following the presented approach is minimized in some sense. In view of our discussions from Section 2.3 and Section 2.4, a good choice should lead to a small number of non-admissible vector fields that have admissible Lie bracket representations of low degree and have a "simple" structure, e.g., as in Section 2.3.2.1. Such choices could be obtained following an optimization-based approach; it is a matter of future research to determine meaningful objective functions, especially in the likely situation that these goals conflict.

2.5.2 Accelerated Saddle-Point Dynamics

The highly oscillatory nature of the approximating inputs naturally leads to an undesired oscillating behavior of the trajectories of the distributed approximation. As discussed in Section 2.3 and as visible in the numerical example depicted in Figure 2.9, the effect on the primal variables, which are in most cases the ones one is most interested in, can be reduced by a proper design of the approximating inputs. Another natural remedy to this problem is to make use of filters which we want to briefly discuss in the following. There are different ways of introducing filters in the feedback loop; in the following we concentrate on the situation depicted in Figure 2.11, where only the signals u_z, u_ν, u_λ are modified by means of low-pass filters G_z, G_ν, G_λ, where G_z, G_ν, G_λ are square stable and proper transfer matrices of appropriate dimension. In view of a distributed implementation we restrict ourselves to diagonal transfer matrices; hence the additional filters do not introduce new variables which are not available to an agent in a distributed setting. It is clear that for these modified saddle-point dynamics to converge to the desired saddle-point of the Lagrangian, it is required to design the filters appropriately and a thorough stability analysis is in place. We do not address this here but emphasize that, as long as the filters are "sufficiently fast", similar stability results can be obtained making use of singular perturbation theory. We further note that similar generalized saddle-point dynamics have been considered in Yamashita, Hatanaka, Yamauchi, and Fujita (2018).

As to the distributed approximation of the filtered saddle-point dynamics, only minor modifications are required. In rough words, the non-admissible terms appearing in the filtered saddle-point dynamics take the same form as the ones without a filter but, since the complete state is augmented by the internal states of the filter, they appear in a different component. To illustrate the effect of the additional filters, we reconsider Example 2 and the numerical results from Figure 2.9 with additional first-order low-pass filters. The results are depicted in Figure 2.10, where, apparently, the filters dampen the oscillations compared to Figure 2.9.

Such filtered saddle-point dynamics can also be interpreted as *higher order saddle-point*

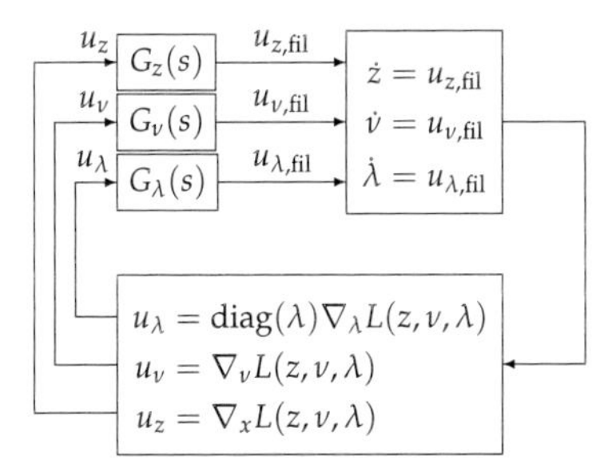

Figure 2.11. Saddle-point dynamics (2.19) with additional low-pass filters G_z, G_ν, G_λ.

dynamics where the minimization in the primal variable as well as the maximization in the dual variables is not performed by means of a standard gradient descent or ascent, respectively, but higher order gradient-based dynamics that can be related to accelerated gradient methods (see, e.g., Michalowsky and Ebenbauer (2014)) are used. In particular, for the solution of constrained optimization problems of the form (2.4), we propose the following generalization of (2.19)

$$\begin{bmatrix} \dot{\mathbf{z}} \\ \dot{\boldsymbol{\nu}} \\ \dot{\boldsymbol{\lambda}} \end{bmatrix} = \begin{bmatrix} A_z \mathbf{z} + B_z \nabla_z L(C_z \mathbf{z}, C_\nu \boldsymbol{\nu}, C_\lambda \boldsymbol{\lambda}) \\ A_\nu \boldsymbol{\nu} + B_\nu \nabla_\nu L(C_z \mathbf{z}, C_\nu \boldsymbol{\nu}, C_\lambda \boldsymbol{\lambda}) \\ A_\lambda \boldsymbol{\lambda} + B_\lambda \mathrm{diag}(\lambda) \nabla_\lambda L(C_z \mathbf{z}, C_\nu \boldsymbol{\nu}, C_\lambda \boldsymbol{\lambda}) \end{bmatrix} \tag{2.122a}$$

$$\begin{bmatrix} z \\ \nu \\ \lambda \end{bmatrix} = \begin{bmatrix} D_z \mathbf{z} \\ D_\nu \boldsymbol{\nu} \\ D_\lambda \boldsymbol{\lambda} \end{bmatrix}, \tag{2.122b}$$

where $\mathbf{z}(t) \in \mathbb{R}^{n_z n}$, $\boldsymbol{\nu}(t) \in \mathbb{R}^{n_\nu n_{\mathrm{eq}}}$, $\boldsymbol{\lambda}(t) \in \mathbb{R}^{n_\lambda n}$, $z(t) \in \mathbb{R}^n$, $\nu(t) \in \mathbb{R}^{n_{\mathrm{eq}}}$, $\lambda(t) \in \mathbb{R}^{n_{\mathrm{ineq}}}$, $n_z, n_\nu, n_\lambda \in \mathbb{N}_{>0}$. For the dynamics (2.122) to converge to a saddle point of the Lagrangian L, i.e., for (2.122) to have an asymptotically stable equilibrium $(\mathbf{z}^\star, \boldsymbol{\nu}^\star, \boldsymbol{\lambda}^\star)$ with $z^\star = D_z \mathbf{z}^\star$, $\nu^\star = D_\nu \boldsymbol{\nu}^\star$, $\lambda^\star = D_\lambda \boldsymbol{\lambda}^\star$, the matrices A_z, A_ν, A_λ, B_z, , B_ν, B_λ, C_z, C_ν, C_λ, D_z, D_ν, D_λ need to be chosen appropriately. We do not provide a solution to this design problem here but only provide some initial ideas to address it. In particular, we propose three different approaches: (i) a robust control reformulation in the framework of integral quadratic constraints (IQCs), (ii) a Lyapunov-based approach, and (iii) a passivity based approach in an input/output setting. We postpone the discussion of (i) to Section 3.5 where we consider similar dynamics in a discrete-time setting; approach (ii) is not discussed in this thesis but can be thought of an extension of Michalowsky and Ebenbauer (2014, 2016) where Lyapunov functions are combined with an S-procedure argumentation. In this subsection, we concentrate on approach (iii). The main idea of this approach is to represent (2.122) as a feedback interconnection of two subsystems, where, roughly speaking, the former is given by the primal dynamics and the latter by the dual dynamics. If we can ensure certain passivity properties of the primal and the dual dynamics, then we can infer the desired closed-loop properties from standard results. We make this more precise in the following. We only consider equality constraints here, i.e., $n_{\mathrm{ineq}} = 0$ and $\boldsymbol{\lambda}$ is neglected in (2.122); we note that inequality constraints can be handled utilizing barrier functions, see Section 3.5.1 for details. Neglecting $\boldsymbol{\lambda}$, for the dynamics (2.122) to have an equilibrium $(\mathbf{z}^\star, \boldsymbol{\nu}^\star)$ with $z^\star = D_z \mathbf{z}^\star$, $\nu^\star = D_\nu \boldsymbol{\nu}^\star$, the matrices A_z, A_ν, B_z, B_ν, C_z, C_ν, D_z, D_ν must adhere to certain conditions. More precisely, we require that there exist matrices $D_z^\dagger \in \mathbb{R}^{n_z n \times n}$, $D_\nu^\dagger \in \mathbb{R}^{n_\nu n_{\mathrm{eq}} \times n_{\mathrm{eq}}}$ such that

$$D_z D_z^\dagger = I_n \qquad D_\nu D_\nu^\dagger = I_{n_{\mathrm{eq}}} \tag{2.123a}$$

$$C_z D_z^\dagger = I_n \qquad C_\nu D_\nu^\dagger = I_{n_{\mathrm{eq}}} \tag{2.123b}$$

$$A_z D_z^\dagger = 0_{n_z n \times n} \qquad A_\nu D_\nu^\dagger = 0_{n_\nu n_{\mathrm{eq}} \times n_{\mathrm{eq}}}. \tag{2.123c}$$

As it turns out, these conditions are sufficient; it is an open question whether, similar to Theorem 3, they are also necessary. We do not elaborate on that here and simply assume

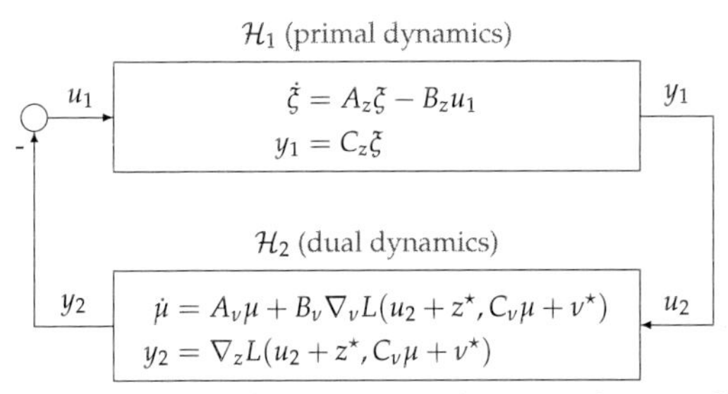

Figure 2.12. Block diagram of (2.124a) separated into two subsystems $\mathcal{H}_1, \mathcal{H}_2$.

that such matrices $D_z^\dagger, D_z^\dagger$ are given. Consider (2.122) together with the state transformation $\xi = \mathbf{z} - D_z^\dagger z^\star$, $\mu = \boldsymbol{\nu} - D_\nu^\dagger \nu^\star$. The transformed dynamics then read as

$$\begin{bmatrix} \dot{\xi} \\ \dot{\mu} \end{bmatrix} = \begin{bmatrix} A_z\xi + B_z\nabla_z L(C_z\xi + z^\star, C_\nu\mu + \nu^\star) \\ A_\nu\mu + B_\nu\nabla_\nu L(C_z\xi + z^\star, C_\nu\mu + \nu^\star) \end{bmatrix} \tag{2.124a}$$

$$\begin{bmatrix} z \\ \nu \end{bmatrix} = \begin{bmatrix} D_z\xi + z^\star \\ D_\nu\mu + \nu^\star \end{bmatrix}. \tag{2.124b}$$

We note that (2.124) has an equilibrium at the origin at which $z = z^\star$, $\nu = \nu^\star$. We then separate (2.124a) into two subsystems $\mathcal{H}_1, \mathcal{H}_2$ as depicted in Figure 2.12. In the following, we only briefly sketch one line of argumentation utilizing this input-output viewpoint. Suppose that $\mathcal{H}_1$ is passive and $\mathcal{H}_2$ is input strictly passive (see, e.g., Khalil (2002, Def. 6.3)), i.e., there exist positive semi-definite storage functions $V_1 : \mathbb{R}^{n_x n} \to \mathbb{R}$, $V_2 : \mathbb{R}^{n_\gamma n_{\mathrm{eq}}} \to \mathbb{R}$, $V_1, V_2 \in \mathcal{C}^1$, such that

$$u_1^\top y_1 \geq \tfrac{\partial V_1}{\partial \xi}(\xi)\dot{\xi}, \qquad u_2^\top y_2 \geq \tfrac{\partial V_2}{\partial \mu}(\mu)\dot{\mu} + \varepsilon u_2^\top u_2 \tag{2.125}$$

for some $\varepsilon > 0$, for all $\xi \in \mathbb{R}^{n_x n}$, $\mu \in \mathbb{R}^{n_\gamma n_{\mathrm{eq}}}$. In the closed loop depicted in Figure 2.12, we then obtain with $V(\xi, \mu) = V_1(\xi) + V_2(\mu)$

$$\dot{V}(\xi, \mu) = \tfrac{\partial V_1}{\partial \xi}(\xi)\dot{\xi} + \tfrac{\partial V_2}{\partial \mu}(\mu)\dot{\mu} \leq -\varepsilon\|y_1\|^2. \tag{2.126}$$

Hence, by LaSalle's Theorem, $y_1(t) = C_z\xi(t) = C_z\mathbf{z}(t) - z^\star$ converges to zero, and we infer that $C_z\xi(t)$ converges to the minimizer $z^\star$ as t tends to infinity. Let $D_z = C_z$. Then $z(t)$ converges to $z^\star$ as t tends to infinity as desired. We hence need to design A_z, B_z, C_z such that $\mathcal{H}_1$ is passive and A_ν, B_ν, C_ν such that $\mathcal{H}_2$ is input strictly passive. By the Positive Real Lemma (see, e.g., Khalil (2002, Lemma 6.2)), given that the pairs (A_z, B_z), (A_ν, B_ν) are controllable and (A_z, C_z), (A_ν, C_ν) are observable, passivity of the linear system $\mathcal{H}_1$ is equivalent to the solvability of a matrix inequality in A_z, B_z, C_z. Similarly, it can be shown that (see Appendix C.1.2), under the additional assumption that H is m-strongly convex for some $m > 0$, input strict passivity of $\mathcal{H}_2$ is implied by passivity of the linear system $(A_v, B_v, C_v, 0)$, which again can be formulated in terms of a matrix inequality in A_v, B_v, C_v, see Appendix C.1.2. This provides the ingredients for analysis and – fixing some of the design parameters – for synthesis of higher order saddle-point dynamics (2.124). We emphasize that this is only a first step and further research needs to be done.

The dynamics (2.122) also have applications to (model-based) Extremum Control problems (Michalowsky & Ebenbauer, 2016). In Extremum Control (Åström & Wittenmark, 1995; Sternby, 1980), the goal is to design a controller for a given dynamic system such that the closed-loop dynamics converge to a setpoint that is specified as the minimizer of some optimization problem. We elaborate on this viewpoint in the subsequent section.

2.5.3 Distributed Extremum (Seeking) Control

There are many practical applications where it is desired to regulate a dynamical system to a setpoint that is optimal with respect to some performance criterion. Examples are, amongst others, power output maximization for power plants or fuel consumption minimization when driving with constant velocity. Likewise, control problems formulated in terms of artificial potential functions such as consensus, obstacle avoidance or formation control fall into that class. Many of the aforementioned problems share the property that the optimal setpoint cannot be determined a priori or changes with time such that common setpoint stabilization techniques do not apply. It is rather required to stabilize or track an unknown extremum which is known as *Extremum Control* (Åström & Wittenmark, 1995; Sternby, 1980). Although originally not limited to such type of problems, by now the term often implies the limitation that not the gradient of the objective function but only the function itself can be evaluated. This is usually referred to as *Extremum Seeking (Control)* (Ariyur and Krstić (2003)). When the underlying dynamic system consists of several smaller units having only local information available, e.g., in optimal power dispatch problem for energy networks or in multi-agent formation control, distributed solutions of such Extremum (Seeking) control problems have been investigated. However, the results available in literature (Dürr et al., 2013; Guay, Vandermeulen, Dougherty, & Mclellan, 2015; C. Li, Qu, & Weitnauer, 2015) require strong assumptions either on the structure of the optimization problem or the communication network. In the following we briefly illustrate how the results of the present chapter can be employed for addressing such distributed Extremum Seeking problems. These ideas have also partly been presented in Michalowsky et al. (2017a).

We next formalize the problem. Consider a multi-agent system consisting of n agents with linear dynamics

$$\dot{x}_i = Ax_i + Bu_i \tag{2.127a}$$

$$y_i = Cx_i \tag{2.127b}$$

where $x_i(t) \in \mathbb{R}^{n_x}$, $u_i(t) \in \mathbb{R}$, $y_i(t) \in \mathbb{R}$, $n_x \in \mathbb{N}_{>0}$, $i = 1, 2, \ldots, n$. The goal in constrained Extremum Control is then to design a controller that generates inputs u_i such that the collection of the agents' outputs $y(t) = [y_i(t)]_{i=1,2,\ldots,n}$ converges to a minimizer of the optimization problem (2.4) as t tends to infinity for all H fulfilling Assumption 1. For scalar, integrator type dynamics, i.e., $A = 0$, $B = 1$, $C = 1$, $n_x = n_u = 1$, a solution to this problem is given by saddle-point dynamics as introduced in Section 2.2. More precisely, the dynamics (2.127) in closed loop with the controller

$$u_i = -e_i^\top \left(\nabla H(x) + \tfrac{\partial a}{\partial x}(x)^\top \nu + \tfrac{\partial c}{\partial x}(x)^\top \lambda\right), \qquad i = 1,2,\dots,n, \tag{2.128a}$$

$$\dot{\nu} = a(x) \tag{2.128b}$$

$$\dot{\lambda} = \operatorname{diag}(\lambda)c(x), \tag{2.128c}$$

where $x = \begin{bmatrix} x_1 & x_2 & \dots & x_n \end{bmatrix}^\top$, yield the saddle-point dynamics (2.19) which we have shown to converge to a solution of (2.4) in Theorem 1. As discussed in detail in the present chapter, this controller is in general not distributed in the sense that each agent only uses locally available variables in its control law. However, we might still employ the very same methodology to obtain distributed control laws; hence our approach also allows us to design distributed controllers for constrained Extremum Control problems for integrator type agents.

In the same manner, utilizing ideas from Dürr et al. (2013), it is then straightforward to design (distributed) Extremum Seeking controllers, i.e., gradient-free variants of (2.128) for this type of problems. This is due to the observation that the gradient of the objective function can as well be represented as a Lie bracket of vector fields that only involve the objective function itself, but not its gradient. More precisely, it has been observed in Dürr et al. (2013) that

$$\nabla H(z) = \sum_{i=1}^{p} [e_i, He_i](z) \tag{2.129}$$

for all $z \in \mathbb{R}^p$. Adapting the wording introduced for the distributed setup, we have represented a gradient-based (viz. non-admissible) vector field by means of Lie bracket of gradient-free (viz. admissible) vector fields. Thus, in the same manner as we have derived distributed algorithms, Lie bracket approximation techniques as explained in Section 2.4 then provide us with the necessary tools to derive gradient-free controllers. We note that (2.129) is not the only possible representation; in fact, a whole class of vector fields has been characterized (Grushkovskaya et al., 2018). In terms of a distributed implementation, depending on the objective function H and the communication graph $\mathcal{G}$, the vector fields He_i in (2.129) can be admissible or not. In the second case, we can employ the techniques developed in Section 2.3 to derive admissible Lie bracket representations of He_i. Note that, if H is additively separable, i.e., $H(z) = \sum_{i=1}^{p} H_i(z_i)$, $H_i : \mathbb{R} \to \mathbb{R}$, then

$$\nabla H(z) = \sum_{i=1}^{p} [e_i, H_i e_i](z) \tag{2.130}$$

and the vector fields $H_i e_i$, $i = 1,2,\dots,p$, are admissible.

We only briefly discuss how the former ideas could be extended to general linear agent dynamics (2.127). The idea to address this problem is to employ higher order saddle-point dynamics (2.122). To explain the idea, we assume that the dynamics (2.127) are given in controller canonical form, i.e.,

$$A = \begin{bmatrix} 0 & 1 & 0 & \dots & 0 \\ 0 & 0 & 1 & \dots & 0 \\ \vdots & & \ddots & \ddots & \\ \vdots & & & & 1 \\ a_1 & a_2 & \dots & \dots & a_{n_x} \end{bmatrix}, \quad B = \begin{bmatrix} 0 \\ 0 \\ \vdots \\ 0 \\ 1 \end{bmatrix}, \tag{2.131}$$

where $a_i \in \mathbb{R}$, $i = 1, 2, \dots, n_x$. We note that any linear dynamics (2.127) can be transformed into controller canonical form given that the pair (A, B) is controllable; hence this assumption is only a mild restriction. Following ideas from Michalowsky and Ebenbauer (2016), we then let the controller be given by

$$u_i = -\sum_{i=2}^{n_x} K_i x_i - K_1 e_i \nabla_x L(C_x x, v, \lambda) \tag{2.132a}$$

$$\dot{v} = \nabla_v L(z, v, \lambda) \tag{2.132b}$$

$$\dot{\lambda} = \operatorname{diag}(\lambda) \nabla_\lambda L(z, v, \lambda), \tag{2.132c}$$

where $K_i \in \mathbb{R}$, $i = 1, 2, \dots, n_x$, $C_x \in \mathbb{R}^{n \times nn_x}$ are controller parameters to be designed. We note that the closed loop of the agent dynamics (2.127) and the controller (2.132) is a particular instance of (2.122), we may hence employ similar methods for controller synthesis. With regard to the scope of this thesis, we do not further elaborate on that here.

2.6 Summary and Conclusion

In this chapter, we established a systematic methodology for the design of distributed optimization algorithms for a large class of convex optimization problems under mild assumptions on the communication topology and the optimization problem. Based on a novel application of Lie bracket averaging techniques, we developed a procedure allowing to systematically generate distributed optimization algorithms from given non-distributed ones. This procedure can be divided into two main steps: (1) finding admissible Lie bracket representations of the non-distributed parts in the non-distributed algorithm and (2) designing distributed approximations of the non-distributed algorithm based on this novel representation. We discussed both steps in detail concentrating on saddle-point dynamics as a prototype for a, in general situations, non-distributed optimization algorithm. Additionally, for Step 1, we also discussed how admissible Lie bracket representations can be found in general situations providing the basics to apply the procedure to a wide range of non-distributed optimization algorithms.

While we showed that the approach can in principle be applied in quite general situations, a major limitation still is the design of suitable approximating sequences that enable the distributed approximation in Step 2 and further research should be carried out into that direction. First, the computation of the parameters of these sequences requires global information and is hence not distributed; although this can be done off-line in a preliminary step, still a distributed implementation, if possible, would be preferable. Second, in practical applications where the magnitude of the sequence parameter is limited, the performance of the distributed approximation heavily depends on the approximating input sequences. Besides addressing this issue in a direct manner by modifying or optimizing the proposed procedure of computing these sequences, it is also worthwhile to consider an indirect approach, i.e., appropriately modify the non-distributed algorithm to be approximated in a distributed fashion. We presented preliminary ideas concerning this approach in Section 2.5.2, however, a systematic design procedure is still lacking and not within the scope of the present thesis. Still, the framework for designing discrete-time optimization algorithms presented in the following chapter could serve as a basis for developing such methods for saddle-point dynamics in particular; we give some more details in Section 3.5.3. As mentioned beforehand, our approach is not limited to saddle-point dynamics and can in principle be utilized to derive distributed approximations of various continuous-time algorithms. Further research should be carried out to determine applications where one could most benefit from the derived methodology. It is also an interesting question whether a similar framework can be established for discrete-time optimization algorithms. Preliminary works on discrete-time Lie bracket averaging techniques (Feiling, Zeller, & Ebenbauer, 2018) give reason to hope that this could indeed be possible.

3

Design of Robust & Structure Exploiting Optimization Algorithms

Fast and reliable optimization algorithms are of key importance in science and engineering. First order, i.e., gradient-based algorithms, are an important subclass that have proved useful in a wide range of applications. In recent years, such algorithms have regained interest since they are particularly suitable for large-scale optimization, which has become more and more important due to its importance in machine learning and data-based applications. In such applications, an optimization algorithm should not only guarantee fast convergence to optimizers but also be robust with respect to noise in the data. While many variants of gradient-based optimization algorithms that yield fast convergence are known in the literature, the design of optimization algorithms that also provide desired robustness properties has not been addressed. Indeed, it has been observed for several known algorithms that there is a trade-off between fast convergence and robustness. However, no framework for systematically analyzing algorithms with regard to these properties or designing algorithms with desired properties exists. Still, a lot of these algorithms fall into the class of *Lur'e systems* (Lur'e & Postnikov, 1944), i.e., a known or to be designed linear system in feedback with an unknown nonlinearity, given by the gradient in the problem at hand. Lur'e systems and the corresponding absolute stability problem are classical control problems leading to celebrated results such as the Popov or the circle criterion, which can be seen as a pioneering contribution to robust control theory. Today's robust control theory provides a quite mature set of tools for analysis and design of controllers and allows to systematically consider various kinds of disturbances. As already mentioned in the introduction, this relation of absolute stability and robust control theory to optimization algorithm analysis and design has not yet been exploited heavily.

In this chapter, we employ this systems theoretic viewpoint on optimization algorithms and provide a systematic framework to the analysis and design of robust and structure exploiting optimization algorithms. In particular, we address the following problem: Given a class of objective functions, design a gradient-based optimization algorithm with a guaranteed convergence rate that also fulfills certain H_2-performance specifications. We show that

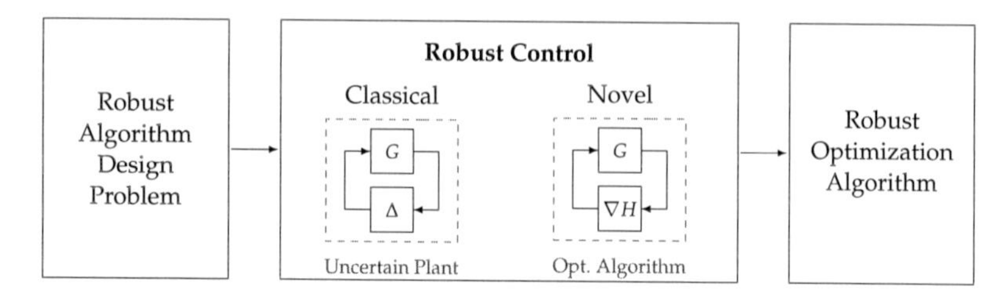

Figure 3.1. An illustration of the general idea of the proposed approach. Interpreting the gradient of the objective function as uncertainty, we reformulate the algorithm design problem as a robust controller synthesis problem and utilize tools from robust control to design robust optimization algorithms for classes of optimization problems.

these performance specifications can be related to noise rejection properties of the algorithm such as the effect of additive gradient noise on the average variance of the algorithm's output. We further address the problem of how to incorporate possible structural properties of the class of objective functions in our framework, and how this can be exploited to design novel optimization algorithms superior to standard ones in terms of convergence rates. To this end, in the spirit of Lessard et al. (2016); Michalowsky and Ebenbauer (2014), we reformulate this design problem as a robust controller synthesis problem and employ integral quadratic constraints theory (IQC theory), see Figure 3.1 for an illustration of the general idea. By building upon and extending these well-established results, we are able to provide a general framework for algorithm analysis and design. To validate the practical applicability and relevance, we provide several numerical results illustrating our methodology. We note that the results presented in this chapter also have been submitted for publication to a large extent, see Michalowsky, Scherer, and Ebenbauer (2020).

Background and Related Work. As mentioned beforehand, at its core the considered problem is an absolute stability problem which is a classic control problem and has led to many famous results both in continuous- (Popov, 1961; Zames & Falb, 1968) and discrete-time (Jury & Lee, 1964; O'Shea & Younis, 1967; Tsypkin, 1964; Willems & Brockett, 1968), culminating in robust control theory, in particular integral quadratic constraints (IQCs), see Megretski and Rantzer (1997) and references therein. However, the idea of utilizing robust control theory to analyze and design first-order optimization algorithms has only come up recently. The probably first work into that direction is Michalowsky and Ebenbauer (2014), where the authors propose a novel class of gradient-based continuous-time optimization algorithms based on a reformulation of the algorithm design problem as a robust state feedback problem with the gradient as the uncertainty and employing simple sector bounds. In the last years, several articles have been published that follow similar ideas. In particular, since the approach is most closely related to the present thesis, we mention Lessard et al. (2016), where the authors as well reformulate the problem as a robust analysis problem and employ an appropriate modification of IQC theory (Boczar, Lessard, Packard, & Recht, 2017; Boczar et al., 2015; Hu & Seiler, 2016) to develop a framework for analyzing the convergence rates of a class of gradient-based optimization algorithms. Since then, several authors have applied (Safavi et al., 2018; Van Scoy et al., 2018) and extended (Fazlyab, Ribeiro, et al., 2018; Freeman, 2018; Zhang, Seiler, & Carrasco, 2019) this framework. However, most of the literature concentrates on convergence rate analysis. Quite recently, the design problem has been addressed (Fazlyab, Morari, & Preciado, 2018; Lessard & Seiler, 2019) and other performance metrics such as robustness have also been studied (Aybat et al., 2019; Cyrus et al., 2018). Still, none of the latter references has exploited the tools inherently provided by the robust control approach. This is in contrast to the work Michalowsky, Scherer, and Ebenbauer (2020), which the present chapter is mainly based on. Therein, the authors embed the algorithm analysis and design problem into the framework of integral quadratic constraints in such a way that many of the existing results from robust control theory can directly be incorporated in a systematic manner.

Apart from a robust control approach, several authors also investigated the problem from a dynamical systems perspective. As one approach that led to several results contributing to a deeper understanding of accelerated gradient methods, discrete-time algorithms were represented as continuous-time dynamic systems together with an appropriate numerical integration scheme (Su et al., 2014; Wibisono et al., 2016). The analysis is then based on the well-established theory for continuous-time dynamic systems. As another direction driven by systems theoretic ideas, Lyapunov theory has been used, both for discrete- and continuous-time dynamics (Wilson, Recht, & Jordan, 2016).

Besides the idea of a systems and control theoretic reformulation, also other approaches have been proposed in the literature for establishing a unified framework for analyzing and designing first-order optimization algorithms. Many of them rely on a proper reformulation as an optimization problem (Drori & Teboulle, 2014; Taylor, Hendrickx, & Glineur, 2017) and the focus lies on algorithm analysis while the design problem is hardly addressed (Drori & Taylor, 2018).

3.1 Problem Formulation

In this chapter, we consider unconstrained convex optimization problems

$$\underset{z \in \mathbb{R}^p}{\text{minimize}} \quad H(z), \tag{3.1}$$

where $H : \mathbb{R}^p \to \mathbb{R}$, $H \in \mathcal{C}^1$, H is strongly convex with convexity modulus $m > 0$ and has Lipschitz continuous gradient with parameter $L \geq m$, i.e., H is in the class $\mathcal{S}_{m,L}$ as defined in Definition 15. Observe that, under these assumptions, for any fixed H in the class, (3.1) has a unique global minimizer which will be denoted by $z^\star$ in the following.

Our goal is to analyze and design gradient-based optimization algorithms that converge to that minimizer $z^\star$. In particular, we consider optimization algorithms of the form

$$x_{t+1} = Ax_t + B\nabla H(Cx_t) \tag{3.2a}$$
$$z_t = Dx_t, \tag{3.2b}$$

where $x_t = \begin{bmatrix} x_{t,1}^\top & \dots & x_{t,n}^\top \end{bmatrix}^\top \in \mathbb{R}^{np}$, $x_{t,i} \in \mathbb{R}^p$, $i \in \{1, \dots, n\}$, $z_t \in \mathbb{R}^p$. More formally, the design problem we want to address is then as follows: Given the objective function parameters $L \geq m > 0$, the dimensions $n, p \geq 1$, and a convergence rate bound $\rho \in (0, 1)$, find matrices $A \in \mathbb{R}^{np \times np}$, $B \in \mathbb{R}^{np \times p}$, $C \in \mathbb{R}^{p \times np}$, $D \in \mathbb{R}^{p \times np}$, such that, for all $H \in \mathcal{S}_{m,L}$, (3.2) has a unique globally asymptotically stable equilbrium at $x^\star$ with $Dx^\star = z^\star$ and there exists $\eta > 0$ such that

$$\|z_t - z^\star\|^2 \leq \eta\rho^{2t}\|z_0 - z^\star\|^2 \tag{3.3}$$

for each $x_0 \in \mathbb{R}^{np}$ with $z_0 := Dx_0$, i.e., z_t converges exponentially with rate ρ to $z^\star$. Note that (3.2) captures several popular optimization algorithms with constant step size such as Gradient Descent, the Heavy Ball Method, or Nesterov's Method, see also (3.67) and Table 3.1. By solving the design problem, we also get a solution to the corresponding analysis problem for these algorithms as a by-product, in which we aim to find an as tight as possible convergence rate bound ρ. We emphasize that (3.2) is an overparametrization of the class of algorithms; in fact, without loss of generality, we can fix $C = \begin{bmatrix} C_1 & C_2 & \dots & C_n \end{bmatrix}$ arbitrarily as long as C_i is non-singular for some $i \in \{1, 2, \dots, n\}$. In particular, we often let $C = \begin{bmatrix} I & 0 & \dots & 0 \end{bmatrix}$.

While convergence rate bounds are an important performance measure, also other performance specifications are key in efficiently solving optimization problems, e.g., how well an algorithm performs in the presence of noise. In this vein, we consider (3.2) together with an additional performance channel and address the extended design problem aiming not only for minimizing the convergence rate but also the bounds on the additional performance channel. We formalize this problem in Section 3.2 and discuss relevant performance specifications for the problem at hand in the following sections.

We are further interested in designing tailored algorithms for certain subclasses of the class of objective functions $\mathcal{S}_{m,L}$ with additional structural properties such as having a diagonal Hessian. Summing up, in rough words our goal is to design gradient-based algorithms of the form (3.2) that are fast, robust and possibly exploit additional structural properties of the objective function.

3.2 Reformulation in the Robust Control Framework

Our approach relies on reformulating the problem as a robust control problem by interpreting the gradient of the objective function H as the uncertainty. The reformulation as well as the adaptation to the specific problem at hand is the main subject of this section.

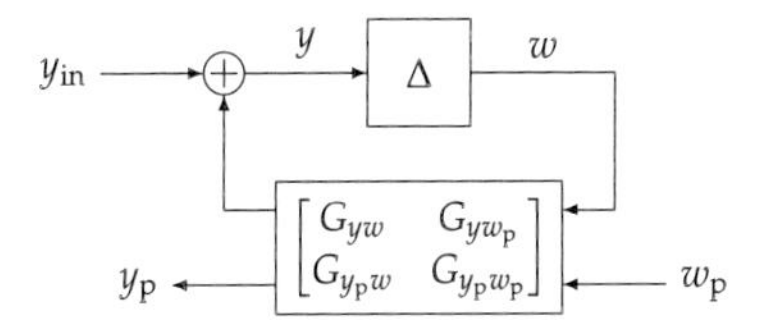

Figure 3.2. The considered feedback interconnection.

3.2.1 Equilibrium Conditions

Before we proceed, we first discuss which properties an algorithm of the form (3.2) must fulfill to be, in principle, a candidate for the previously posed problems. This important question is answered in the following result which provides necessary and sufficient conditions for an algorithm of the form (3.2) to possess an equilibrium at the global minimizer of H for all $H \in \mathcal{S}_{m,L}$.

Theorem 3. Let an algorithm in the form (3.2) described by A, B, C, D be given and assume that the pair (A, C) or the pair (A, D) is observable. Then, (3.2) has an equilibrium $x^\star$ with the property $Dx^\star = Cx^\star = z^\star$ for any $H \in \mathcal{S}_{m,L}$, $L \geq m > 0$ arbitrary but fixed, if and only if there exists $D^\dagger \in \mathbb{R}^{np \times p}$ such that

$$DD^\dagger = I_p \tag{3.4a}$$

$$CD^\dagger = I_p \tag{3.4b}$$

$$(A - I)D^\dagger = 0_{np \times p}. \tag{3.4c}$$

holds. •

A proof of Theorem 3 is provided in Appendix B.2.1. We emphasize that the conditions (3.4) are necessary and sufficient, hence assuming that these conditions hold is no restriction of the class of algorithms. Note that (3.4c) implies that A must have at least p eigenvalues at one. The proof also reveals that $x^\star = D^\dagger z^\star$ is the unique equilibrium of (3.2) with the desired property. We further note that, under the assumption that D has full rank, the convergence rate bound (3.3) holds if and only if the full state x_k itself converges exponentially with rate ρ to $D^\dagger z^\star$. In view of this, the choice of D does not alter the achievable convergence rate. Hence, remembering the previous discussion that (3.2) is an overparametrization, we often let $D = C = \begin{bmatrix} I & 0 & \dots & 0 \end{bmatrix}$, $D^\dagger = D^\top$, such that (3.4a), (3.4b) are ensured to hold. In the remainder of this chapter we limit ourselves to algorithms described by A, B, C, D that fulfill all conditions in Theorem 3.

3.2.2 Introduction to Integral Quadratic Constraints

In the following we briefly present some standard results on robust stability and performance analysis via integral quadratic constraints (IQCs). For a more detailed treatment of the subject, we refer the reader to the classical paper Megretski and Rantzer (1997) as well as Veenman, Scherer, and Köroğlu (2016) in a continuous-time setup. These results literally carry over to the discrete-time setup, see, e.g., Fetzer and Scherer (2017); Kao (2012). In the following, we consider feedback interconnections of the form (cf. Figure 3.2)

$$y = G_{yw}w + G_{yw_p}w_p + y_{in} \tag{3.5a}$$

$$y_p = G_{y_pw}w + G_{y_pw_p}w_p \tag{3.5b}$$

$$w = \Delta(y), \tag{3.5c}$$

consisting of a linear system with stable transfer matrix $G_{yw} \in \mathcal{RH}_\infty^{n_y \times n_w}$ in feedback with an uncertainty $\Delta : \ell_2^{n_y} \to \ell_2^{n_w}$ and an additional performance channel from w_p to y_p, where $G_{yw_p} \in \mathcal{RH}_\infty^{n_y \times n_{w_p}}$, $G_{y_pw} \in \mathcal{RH}_\infty^{n_{y_p} \times n_w}$ as well as $G_{y_pw_p} \in \mathcal{RH}_\infty^{n_{y_p} \times n_{w_p}}$. Let $\mathbf{\Delta} \subset \{\Delta \in \mathcal{L}(\ell_2^{n_y}, \ell_2^{n_w}), \Delta \text{ causal}\}$ denote the set of uncertainties, i.e., $\Delta \in \mathbf{\Delta}$. It is common to impose the following assumption:

Assumption 4. If $\Delta \in \mathbf{\Delta}$ then $\tau\Delta \in \mathbf{\Delta}$ for all $\tau \in [0, 1]$. •

In many cases, this assumption can be ensured by a proper redefinition of the uncertainty. In particular, this assumption will be met by the class of uncertainties we will consider in our problem setup. Our goal is to show robust stability of the feedback interconnection without the performance channel for all $\Delta \in \mathbf{\Delta}$ and robust performance concerning the performance channel from w_p to y_p for all $\Delta \in \mathbf{\Delta}$ with respect to some (integral) quadratic performance criterion. In the remainder of this section, we briefly repeat the basic definitions and results from robust control and IQC theory. We first give a proper definition of robust stability.

Definition 6 (Robust stability). The feedback interconnection (3.5) is said to be *robustly stable against* $\mathbf{\Delta}$ if, for $w_p = 0$, it is well-posed and the ℓ_2-gain of the map from y_{in} to y is bounded for all $\Delta \in \mathbf{\Delta}$. •

Therein, well-posedness of the feedback interconnection means that the map $q \mapsto (I - G_{yw}\Delta)q$ has a causal inverse for all $\Delta \in \mathbf{\Delta}$. We next give the definition of an integral quadratic constraint (IQC).

Definition 7 (IQC). Let a so-called multiplier $\Pi : \mathcal{RL}_\infty^{(p_1+p_2)\times(p_1+p_2)}$, $p_1, p_2 \in \mathbb{N}_{>0}$, be given. We say that *two signals* $q_1 \in \ell_2^{p_1}, q_2 \in \ell_2^{p_2}$ *satisfy the IQC defined by* Π if

$$\mathrm{IQC}(\Pi, q_1, q_2) = \int_0^{2\pi} \begin{bmatrix} \widehat{q_1}(e^{i\omega}) \\ \widehat{q_2}(e^{i\omega}) \end{bmatrix}^* \Pi(e^{i\omega}) \begin{bmatrix} \widehat{q_1}(e^{i\omega}) \\ \widehat{q_2}(e^{i\omega}) \end{bmatrix} \mathrm{d}\omega \geq 0. \tag{3.6}$$

We further say that *an operator* $\Delta : \ell_2^{p_1} \to \ell_2^{p_2}$ *satisfies the IQC defined by* Π if

$$\mathrm{IQC}(\Pi, q_1, \Delta(q_1)) \geq 0 \tag{3.7}$$

holds for all $q_1 \in \ell_2^{p_1}$. •

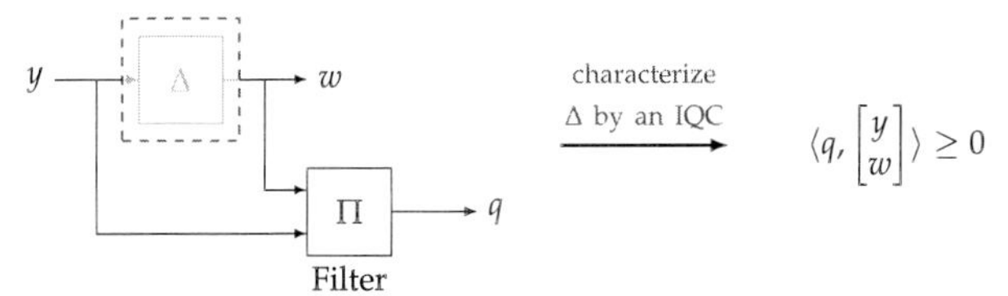

Figure 3.3. An illustration of the idea of an IQC. An unknown troublesome component Δ is replaced by an inequality on the inner product of the components input/output and a filtered signal q.

Remark 6. By Parseval's Theorem, the IQC (3.6) can equivalently be formulated in the time domain as

$$\left\langle \begin{bmatrix} q_1 \\ q_2 \end{bmatrix}, \phi_\Pi \begin{bmatrix} q_1 \\ q_2 \end{bmatrix} \right\rangle \geq 0, \tag{3.8}$$

where ϕ_Π is the linear operator defined by Π as introduced in (A.16). •

In general, IQCs can be used to i) characterize the uncertain operator Δ (see Figure 3.3 for an illustration) as well as to ii) describe performance criteria. For the case i), we try to find a set of suitable multipliers $\mathbf{\Pi}$ such that each $\Delta \in \mathbf{\Delta}$ satisfies the IQC defined by Π for all $\Pi \in \mathbf{\Pi}$. Similarly, for ii), we let $q_1 = w_\mathrm{p}$, $q_2 = y_\mathrm{p}$ and specify a set of multipliers $\mathbf{\Pi}_p \subset \mathcal{RL}_\infty^{(n_{w_\mathrm{p}}+n_{y_\mathrm{p}})\times(n_{w_\mathrm{p}}+n_{y_\mathrm{p}})}$ that characterize the desired performance criterion imposed on the performance channel. The corresponding performance IQC is then given by

$$\mathrm{IQC}(-\Pi_\mathrm{p}, w_\mathrm{p}, y_\mathrm{p}) \geq 0, \tag{3.9}$$

where $\Pi_\mathrm{p} \in \mathbf{\Pi}_\mathrm{p}$. We then define robust performance of (3.5) as follows.

Definition 8 (Robust performance). We say that the feedback interconnection (3.5) *achieves robust performance with respect to* Π_p *against* $\mathbf{\Delta}$ if the feedback interconnection is robustly stable against $\mathbf{\Delta}$ and, for $y_\mathrm{in} = 0$, (3.9) holds for $\Pi_\mathrm{p} \in \mathbf{\Pi}_\mathrm{p}$ and all $\Delta \in \mathbf{\Delta}$. •

In the following we limit ourselves to performance multipliers where the block corresponding to the quadratic terms in y_p is positive semi-definite, i.e., we assume

$$\mathbf{\Pi}_\mathrm{p} \subset \left\{ \begin{bmatrix} \Pi_{\mathrm{p},11} & \Pi_{\mathrm{p},12} \\ \Pi_{\mathrm{p},12}^* & \Pi_{\mathrm{p},22} \end{bmatrix} \in \mathcal{RL}_\infty^{(n_{w_\mathrm{p}}+n_{y_\mathrm{p}})\times(n_{w_\mathrm{p}}+n_{y_\mathrm{p}})} \mid \Pi_{\mathrm{p},22} \overset{\mathbb{T}}{\succeq} 0 \right\}. \tag{3.10}$$

This assumption is met by the most relevant performance criteria and, in particular, it holds for all performance criteria we consider in the present manuscript. We next state the well-known IQC stability theorem.

Theorem 4. Consider the feedback interconnection (3.5). Let some set of multipliers $\mathbf{\Pi} \subset \mathcal{RL}_\infty^{(n_y+n_w)\times(n_y+n_w)}$ as well as some set of performance multipliers $\mathbf{\Pi}_\mathrm{p}$ with (3.10) be given. Assume that G_{yw} has all its poles in the open unit disk and assume that $\Delta \in \mathbf{\Delta}$ where $\mathbf{\Delta}$ fulfills Assumption 4. Suppose further that

1. the interconnection (3.5) is well-posed for all $\Delta \in \mathbf{\Delta}$;

2. each $\Delta \in \mathbf{\Delta}$ satisfies the IQC defined by Π for all $\Pi \in \mathbf{\Pi}$.

If there exist $\Pi \in \mathbf{\Pi}$ and $\Pi_\mathrm{p} \in \mathbf{\Pi}_\mathrm{p}$ such that

$$\begin{bmatrix} \\ \star \\ \\ \end{bmatrix}^* \left[\begin{array}{c|c} \Pi & 0 \\ \hline 0 & \Pi_\mathrm{p} \end{array}\right] \left[\begin{array}{cc} G_{yw} & G_{yw_\mathrm{p}} \\ I & 0 \\ \hline 0 & I \\ G_{y_\mathrm{p}w} & G_{y_\mathrm{p}w_\mathrm{p}} \end{array}\right] \overset{\mathbb{T}}{\prec} 0, \tag{3.11}$$

then the interconnection (3.5) is robustly stable against $\mathbf{\Delta}$ and it achieves robust performance w.r.t. Π_p against $\mathbf{\Delta}$. •

For the sake of completeness, a proof is provided in Appendix B.2.2. The premise in applying Theorem 4 is then to find a class of multipliers valid for the class of uncertainties at hand as well as an IQC formulation of the desired performance criterion such that all assumptions in the latter Theorem are met. In view of a tractable implementation, it is then common to reformulate the frequency domain inequality (FDI) given by (3.11) as a matrix inequality employing the KYP-Lemma (see Lemma 16 in Appendix A.6), thus requiring state-space representations of all transfer functions involved. While standard, we discuss this step for the specific problem in Section 3.2.6.

3.2.3 Problem Reformulation

We next embed the design problem from Section 3.1 in the standard setup as introduced in Section 3.2.2. To this end, consider (3.2) initialized at any x_0 together with the state transformation $\xi_t = x_t - D^\dagger z^\star$, where $D^\dagger \in \mathbb{R}^{np \times p}$ is such that (3.4) hold. The transformed algorithm together with an additional performance channel from w_p to y_p as described in Section 3.1 then takes the form

$$\xi_{t+1} = A\xi_t + B\nabla H(C\xi_t + z^\star) + B_\mathrm{p} w_{\mathrm{p},t} \tag{3.12a}$$

$$z_t = D\xi_t + z^\star \tag{3.12b}$$

$$y_{\mathrm{p},t} = C_\mathrm{p}\xi_t + D_\mathrm{p} w_{\mathrm{p},t}, \tag{3.12c}$$

with initial condition $\xi_0 = x_0 - D^\dagger z^\star$ and where $B_\mathrm{p} \in \mathbb{R}^{np \times n_{w_\mathrm{p}}}$, $C_\mathrm{p} \in \mathbb{R}^{n_{y_\mathrm{p}} \times np}$ and $D_\mathrm{p} \in \mathbb{R}^{n_{y_\mathrm{p}} \times n_{y_\mathrm{p}}}$, $n_{w_\mathrm{p}}, n_{y_\mathrm{p}} \in \mathbb{N}_{>0}$. Note that if ξ converges exponentially to zero with rate ρ, then so does x converge to $D^\dagger z^\star$ in (3.2); in other words (3.3) holds.

As we are aiming for analyzing and designing A, B, C in (3.12) for a whole class of objective functions, we interpret ∇H as an uncertainty and define the causal and bounded uncertain operator $\Delta_H : \ell_e^p \to \ell_e^p$ as

$$\Delta_H(y)_t := \nabla H(y_t + z^\star) - my_t \tag{3.13}$$

for any $y = [y_0, y_1, y_2, \dots] \in \ell_e^p$. The corresponding set of all admissible operators is then defined as

$$\boldsymbol{\Delta}(m, L) := \{\Delta_H : H \in \mathcal{S}_{m,L}\}. \tag{3.14}$$

Note that the set $\boldsymbol{\Delta}(m, L)$ fulfills Assumption 4. Observe further that each $\Delta \in \boldsymbol{\Delta}(m, L)$ is a slope-restricted operator with slope between 0 and $L - m$. In the spirit of Section 3.2.2 and (3.5) we drop the output (3.12b), write (3.12) as

$$\xi_{t+1} = (A + mBC)\xi_t + Bw_t + B_\mathrm{p} w_{\mathrm{p},t} \tag{3.15a}$$
$$y_t = C\xi_t \tag{3.15b}$$
$$y_{\mathrm{p},t} = C_\mathrm{p}\xi_t + D_\mathrm{p} w_{\mathrm{p},t} \tag{3.15c}$$
$$w_t = \Delta_H(y)_t \tag{3.15d}$$

and identify the transfer functions in (3.5) with

$$\begin{bmatrix} G_{yw} & G_{yw_\mathrm{p}} \\ G_{y_\mathrm{p}w_\mathrm{p}} & G_{y_\mathrm{p}w} \end{bmatrix} \sim \left(\begin{array}{c|c} A + mBC & \begin{bmatrix} B & B_\mathrm{p} \end{bmatrix} \\ \hline \begin{bmatrix} C \\ C_\mathrm{p} \end{bmatrix} & \begin{bmatrix} 0 & 0 \\ 0 & D_\mathrm{p} \end{bmatrix} \end{array} \right). \tag{3.16}$$

We note that the latter feedback interconnection is well-posed since G_{yw} is strictly proper and Δ is a static uncertainty. The original design problem can then be formulated as follows: Given a set of operators $\boldsymbol{\Delta}(m, L)$ (or a structured subclass thereof, see Section 3.4) and a performance channel described by $B_\mathrm{p}, C_\mathrm{p}, D_\mathrm{p}$, our goal is to design A, B, C in such a way that (a) the transformed algorithm dynamics (3.15) have a globally asymptotically stable equilibrium at $\xi^\star = 0$ for all $\Delta_H \in \boldsymbol{\Delta}(m, L)$ and ξ converges exponentially to 0 with rate ρ and (b) the performance channel defined by the map from w_p to y_p in (3.12) fulfills a specified performance bound. Note that we do no longer consider D as a design variable since it does not have an effect on the goals (a), (b). Our approach to address the problem at hand is then to make use of IQC theory, in particular Theorem 4. The key steps (S1)–(S3) that pave the way to an implementable solution of the design problem are then as follows:

(S1) Extend Theorem 4 to allow for exponential stability results (Section 3.2.4, in particular Theorem 5).

(S2) Derive IQCs valid for the class of uncertainties that are suitable for the exponential stability result (Section 3.2.5, in particular Theorem 6).

(S3) Determine state-space representations of all transfer functions occurring in the frequency domain inequality (Section 3.2.6).

We note that the last step is more or less standard; still, we include it here in view of our goal of presenting an easily implementable framework.

3.2.4 Exponential Convergence via IQCs

We begin with the first step (S1) of extending Theorem 4 which only allows to conclude ℓ_2-stability results for (3.15). While it is noted in Megretski and Rantzer (1997) that ℓ_2-stability implies exponential stability for some classes of systems, the resulting rate bounds are often conservative (Boczar et al., 2017). In Boczar et al. (2015); Lessard et al. (2016), the concept of ρ-IQCs is introduced to derive exponential stability results. In the present manuscript, we adopt these ideas with a major focus on embedding the concept into the existing IQC framework introduced in Section 3.2.2. As it will get apparent, this allows for a more systematic approach to derive IQCs for robust exponential convergence guarantees from standard IQCs.

We first give a proper definition of robust exponential stability. To this end, we need to step from the input-output framework introduced in Section 3.2.2 to state-space descriptions. Let a state-space representation of (3.5) be given by

$$x_{t+1} = Ax_t + B_1 w_t + B_2 w_{\mathrm{p},t} \tag{3.17a}$$
$$y_t = C_1 x_t + D_{11} w_t + D_{12} w_{\mathrm{p},t} \tag{3.17b}$$
$$y_{\mathrm{p},t} = C_2 x_t + D_{21} w_t + D_{22} w_{\mathrm{p},t} \tag{3.17c}$$
$$w_t = \Delta(y)_t, \tag{3.17d}$$

with initial condition $x_0 \in \mathbb{R}^n$ and $y_{\mathrm{in}} = 0$. Under the assumption that $\Delta(0) = 0$, i.e., the origin is an equilibrium of (3.17) for $w_{\mathrm{p}} = 0$, we then have the following definition:

Definition 9 (Robust exponential stability). We say that the origin of (3.17) is *robustly exponentially stable against* $\mathbf{\Delta}$ *with rate* $\rho \in (0,1)$ if, for $w_{\mathrm{p}} = 0$, it is globally (uniformly) exponentially stable with rate ρ for all $\Delta \in \mathbf{\Delta}$, i.e., there exists $\eta > 0$ such that $\|x_t\|^2 \leq \eta\rho^{2t}\|x_0\|^2$ for all $t \in \mathbb{N}$, for each $x_0 \in \mathbb{R}^n$ and each $\Delta \in \mathbf{\Delta}$. •

Hence, if we can show that the origin is robustly exponentially stable against $\mathbf{\Delta}(m, L)$ as defined in (3.14) with rate $\rho \in (0,1)$ for the transformed dynamics (3.15), then we conclude that the algorithm (3.2) converges with rate ρ to $x^\star = D^\dagger z^\star$ for all $H \in \mathcal{S}_{m,L}$ and (3.3) holds. Since robust exponential stability and additional performance specifications are independent of each other, for the following discussions we neglect the additional performance channel in (3.15).

The idea for obtaining exponential stability results from ℓ_2-stability statements is to make use of a proper time-varying transformation of the signals. Such signal transformations have a long history in convergence rate analysis, e.g., so-called exponential weightings have already been mentioned in Desoer and Vidyasagar (1975) and similar signal transformations are used in Antipin (1994) in the context of optimization. In particular, as in Boczar et al. (2015), for any $\rho \in (0,1]$ and any $p \in \mathbb{N}_{>0}$, we define two operators $\rho_+, \rho_- : \ell_e^p \to \ell_e^p$ as

$$\left(\rho_+(y)\right)_t = \rho^t y_t, \quad \left(\rho_-(y)\right)_t = \rho^{-t} y_t, \tag{3.18}$$

where $t \in \mathbb{N}$. For $\rho \in (0,1)$, we then introduce the signal space

$$\ell_{2,\rho}^p = \{y \in \ell_e^p : \rho_-(y) \in \ell_2^p\} \subset \ell_2^p. \tag{3.19}$$

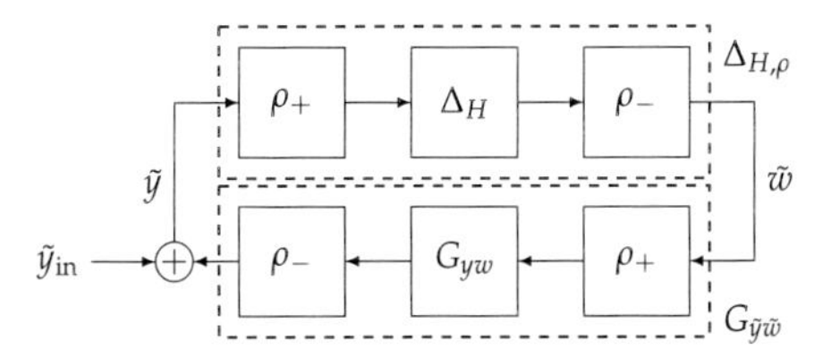

Figure 3.4. Modified feedback interconnection.

Note that any $y \in \ell_{2,\rho}^p$ decays exponentially with rate ρ in the future, i.e., there exists $\eta > 0$ such that $\|y_t\| \leq \eta e^{-\rho k}$ for all $t \in \mathbb{N}$. Note further that

$$\rho_+ : \ell_2^p \to \ell_{2,\rho}^p \qquad \rho_- : \ell_{2,\rho}^p \to \ell_2^p \tag{3.20}$$

are bijective and inverse to each other, i.e., $\rho_+ \circ \rho_- = \mathrm{id}_{\ell_{2,\rho}}$ and $\rho_- \circ \rho_+ = \mathrm{id}_{\ell_2}$. Now, consider (3.15) and suppose that $y, w \in \ell_{2,\rho}^p$ and $\xi \in \ell_{2,\rho}^{np}$ for some $\rho \in (0,1)$. Then $\tilde{y} = \rho_-(y)$, $\tilde{w} = \rho_-(w)$, $\tilde{\xi} = \rho_-(\xi)$ are ℓ_2-signals and we have

$$\tilde{\xi}_{t+1} = \rho^{-1}(A + mBC)\tilde{\xi}_t + \rho^{-1}B\tilde{w}_t \tag{3.21a}$$
$$\tilde{y}_t = C\tilde{\xi}_t \tag{3.21b}$$
$$\tilde{w}_t = \rho^{-t}\Delta_H(\rho^t\tilde{y}_t), \tag{3.21c}$$

with initial condition $\tilde{\xi}_0 = \xi_0$ after neglecting the performance channel. Conversely, if we can show that $\tilde{y}, \tilde{w}, \tilde{\xi}$ in (3.21) reside in ℓ_2, then the original signals y, w, ξ in (3.15) will reside in $\ell_{2,\rho}$ and hence, they decay exponentially. This basically captures the main idea of the transformation since, as we will see next, the latter ℓ_2-property of (3.21) can be shown employing standard IQC methods. However, while seeming technical, one of the key challenges in doing so is to ensure certain boundedness conditions on the operators in both signal spaces ℓ_2 and $\ell_{2,\rho}$.

To this end, observe that by (3.21) the transfer function $G_{\tilde{y}\tilde{w}}$ from $\tilde{w}$ to $\tilde{y}$ is given by

$$G_{\tilde{y}\tilde{w}} \sim \left(\rho^{-1}(A + mBC), \rho^{-1}B, C, 0\right), \tag{3.22}$$

and thus fulfills

$$G_{\tilde{y}\tilde{w}}(z) = G_{yw}(\rho z). \tag{3.23}$$

We may then write (3.21) as

$$\tilde{y} = G_{\tilde{y}\tilde{w}}\tilde{w} \tag{3.24a}$$
$$\tilde{w} = \left(\rho_- \circ \Delta_H \circ \rho_+\right)(\tilde{y}), \tag{3.24b}$$

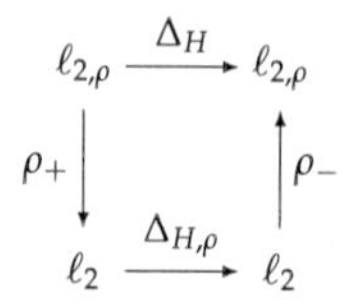

Figure 3.5. A schematic overview of the maps and signal spaces in the transformed loop depicted in Figure 3.4.

see also Figure 3.4. To apply the standard IQC result Theorem 4 to the transformed interconnection, we require the transformed uncertainty

$$\Delta_{H,\rho} = \rho_- \circ \Delta_H \circ \rho_+ \tag{3.25}$$

to be a bounded operator mapping ℓ_2^p to ℓ_2^p. As we show next, this is the case here since Δ_H is bounded, static and time-invariant. To this end, first note that there exists $\beta \in \mathbb{R}_{\geq 0}$ such that $\|\left(\Delta_H(y)\right)_k\| = \|\Delta_H(y_t)\| \leq \beta\|y_t\|$ for any $t \in \mathbb{N}$ and any $y \in \ell_2^p$. With $\ell_{2,\rho}^p \subset \ell_2^p$, we then infer that for any $y \in \ell_{2,\rho}^p$ we have

$$\|\left(\rho_-(\Delta_H(y))\right)_t\| = \rho^{-t}\|\Delta_H(y_t)\| \leq \rho^{-t}\beta\|y_t\|. \tag{3.26}$$

With $y \in \ell_{2,\rho}^p$, i.e., $\sum_{t=0}^{\infty} \rho^{-2t}\|y_t\|^2$ is finite, we conclude that $\rho_-(\Delta(y)) \in \ell_2^p$ for any $y \in \ell_{2,\rho}^p$, hence $\Delta_H : \ell_{2,\rho}^p \to \ell_{2,\rho}^p$ and $\Delta_{H,\rho} : \ell_2^p \to \ell_2^p$ as required, see also Figure 3.5 for an overview of the relations. We next show that $\Delta_{H,\rho}$ is bounded on ℓ_2^p. For $\tilde{y} \in \ell_2^p$, we indeed have $y := \rho_+(\tilde{y}) \in \ell_{2,\rho}$ and thus, using (3.26),

$$\|\Delta_{H,\rho}(\tilde{y})\| = \|\rho_-(\Delta_H(y))\| \leq \beta\|\rho_-(y)\| = \beta\|\tilde{y}\|. \tag{3.27}$$

Consequently, $\Delta_{H,\rho}$ has the required properties, and we define the corresponding set of uncertainties as

$$\mathbf{\Delta}_\rho(m, L) = \left\{\Delta_{H,\rho} = \rho_- \circ \Delta_H \circ \rho_+ \,|\Delta_H \in \mathbf{\Delta}(m, L)\right\}. \tag{3.28}$$

The interconnection (3.24) has the same structure as the standard feedback interconnection (3.5); thus, we may apply Theorem 4, which leads to the following modification as applied to the problem at hand.

Theorem 5. Consider (3.15) and let the transfer functions be defined according to (3.16). Let $L \geq m > 0$ be given and fix $\rho \in (0, 1)$. Suppose that all eigenvalues of $A + mBC$ are located in the open disk of radius ρ. Let some set of multipliers $\mathbf{\Pi}(\rho) \subset \mathcal{RL}_\infty^{2p\times 2p}$ parametrized by ρ and some set of performance multipliers $\mathbf{\Pi}_\mathrm{p}$ in the form (3.10) be given. Suppose that

1. the interconnection depicted in Figure 3.4 is well-posed for all $\Delta_{H,\rho} \in \mathbf{\Delta}_\rho(m, L)$;

2. each $\Delta_{H,\rho} \in \mathbf{\Delta}_\rho(m, L)$ satisfies the IQC defined by Π for all $\Pi \in \mathbf{\Pi}(\rho)$.

If there exist $\Pi_1 \in \mathbf{\Pi}(\rho)$, $\Pi_2 \in \mathbf{\Pi}(1)$ and $\Pi_\mathrm{p} \in \mathbf{\Pi}_\mathrm{p}$ such that

$$\begin{bmatrix} \star \end{bmatrix}^* \Pi_1 \begin{bmatrix} G_{\tilde{y}\tilde{w}} \\ I \end{bmatrix} \overset{\mathrm{T}}{\prec} 0 \tag{3.29a}$$

$$\begin{bmatrix} \\ \star \\ \\ \end{bmatrix}^* \left[\begin{array}{c|c} \Pi_2 & 0 \\ \hline 0 & \Pi_\mathrm{p} \end{array}\right] \left[\begin{array}{cc} G_{yw} & G_{yw_\mathrm{p}} \\ I & 0 \\ \hline 0 & I \\ G_{y_\mathrm{p}w} & G_{y_\mathrm{p}w_\mathrm{p}} \end{array}\right] \overset{\mathrm{T}}{\prec} 0, \tag{3.29b}$$

where $G_{\tilde{y}\tilde{w}}$ is defined in (3.22), then the origin is robustly exponentially stable against $\mathbf{\Delta}(m, L)$ with rate ρ for (3.15) and it achieves robust performance w.r.t. Π_p against $\mathbf{\Delta}(m, L)$. •

A proof can be found in Appendix B.2.3. In terms of the original problem described in Section 3.1, the previous Theorem provides a result that allows the analysis and design of optimization algorithms of the form (3.2) in terms of convergence rates as well as performance. We note that the performance FDI (3.29b) is equivalent to the FDI (3.11) in the classical result Theorem 4 and exponential convergence is captured by the FDI (3.29a). Compared to Theorem 4, we then require an adapted set of IQCs $\mathbf{\Pi}(\rho)$ for the set of transformed uncertainties $\mathbf{\Delta}_\rho(m, L)$; determining such a set will be the main subject of the following section. Following the steps (S1)–(S3), we then also discuss how to reformulate the two FDIs (3.29a), (3.29b) as matrix inequalities, thereby paving the way to an implementable solution.

Remark 7. In principle, the approach is not limited to exponential convergence rates. By employing suitable time-varying transformations, it is expected that similar results can be obtained for other types of convergence rate specifications, for example to obtain polynomial convergence rate guarantees. •

3.2.5 IQCs for the class $\mathcal{S}_{m,L}$

The main goal of this section is to determine a set of IQCs valid for the class of transformed uncertainties $\mathbf{\Delta}_\rho(m, L)$ (step (S2)). In the following we show how to systematically derive such IQCs from classical IQCs for the original set of uncertainties $\mathbf{\Delta}(m, L)$. This is in contrast to the approach from Boczar et al. (2017), Lessard et al. (2016) and allows us to not only recover the modified IQCs from these references but also to easily extend them to the case of anticausal multipliers or to include additional structural properties of the uncertain operator, see Section 3.4. We illustrate the procedure by means of the problem at hand; we emphasize that the same methodology applies in more general situations employing the general result Lemma 7. With $\mathbf{\Delta}(m, L)$ being a set of slope-restricted operators, we can employ Zames-Falb IQCs that have been extensively studied in the literature, see, e.g., Zames and Falb (1968), Willems (1971), Mancera and Safonov (2005). While there exist other related IQCs for slope-restricted operators based on discrete-time variants of the Popov criterion such as the Tsypkin or the Jury-Lee criteria, quite recently it has been shown in Fetzer and Scherer (2017) that, in the discrete-time setup, these are included in the set of Zames-Falb IQCs; thus we concentrate on this class here. We give a brief wrap up of those classical results in the sequel. In the literature, Zames-Falb IQC are commonly stated in the time domain; following Remark 6 this amounts to classifying the corresponding operator ϕ. To this end, we utilize the following definition introduced in Willems and Brockett (1968).

Definition 10 (Doubly hyperdominant matrix). A matrix $M = [m_{ij}]_{i,j\in\{0,1,\dots,r\}}$, $r \in \mathbb{N}$, is said to be *doubly hyperdominant* if

$$m_{ij} \leq 0 \text{ for } i \neq j \text{ and } \sum_{k=0}^{r} m_{kj} \geq 0, \sum_{k=0}^{r} m_{ik} \geq 0 \tag{3.30}$$

for all $i, j \in \{0, 1, \dots, r\}$. •

Let $O_T : \ell_e^p \to \ell_f^p$, $T \in \mathbb{N}$, denote the truncation operator, i.e.,

$$O_T(y) = (y_0, y_1, \dots, y_T, 0, 0, \dots) \tag{3.31}$$

for $y \in \ell_e^p$. An infinite (block) matrix $M = (M_{ij})_{i,j\in\mathbb{N}}$, $M_{ij} \in \mathbb{R}^{p\times p}$, defines a linear operator $M : \ell_f^p \to \ell_e^p$ in a natural fashion; the adjoint operator $M^\top : \ell_f^p \to \ell_e^p$ is then defined by the matrix transpose. Doubly hyperdominance for such operators is then defined as follows:

Definition 11 (Doubly hyperdominant operator). We call a linear operator $M : \ell_f^p \to \ell_e^p$, $p \in \mathbb{N}_{>0}$, *doubly hyperdominant* if the associated matrix of the truncated operator $M_T = O_T M O_T$ is doubly hyperdominant according to Definition 10 for each $T \in \mathbb{N}$. •

Let $\mathbf{M}$ denote the set of all infinite matrices defining doubly hyperdominant operators. We then have the following result; a proof is provided in Appendix B.2.4.

Lemma 6. Let $L \geq m > 0$ be given and let $\mathbf{\Delta}(m, L)$ be defined according to (3.14). Let $M : \ell_f^p \to \ell_e^p$ be a linear operator defined by the infinite matrix $M = \bar{M} \otimes I_p$, $\bar{M} \in \mathbf{M}$. Then, for all $y \in \ell_2^p$, all $\Delta_H \in \mathbf{\Delta}(m, L)$ and all $T \in \mathbb{N}$, we have

$$\left\langle \begin{bmatrix} y \\ \Delta_H(y) \end{bmatrix}, O_T W^\top \begin{bmatrix} 0 & M^\top \\ M & 0 \end{bmatrix} W O_T \begin{bmatrix} y \\ \Delta_H(y) \end{bmatrix} \right\rangle \geq 0 \tag{3.32}$$

with

$$W = \begin{bmatrix} (L-m)\mathrm{id} & -\mathrm{id} \\ 0 & \mathrm{id} \end{bmatrix}. \tag{3.33}$$

•

Remark 8. If $\mathbf{\Delta}(m, L)$ has the additional property that $\Delta(y) = -\Delta(-y)$ for any $y \in \ell_2^p$, $\Delta \in \mathbf{\Delta}(m, L)$, i.e., each Δ is an odd operator, then (3.32) persists to hold for $\bar{M}$ being doubly dominant, see Willems (1971) for a definition. •

We note that if M is a bounded operator, then the limit of (3.32) as $T \to \infty$ is well-defined and the inequality corresponds to a standard time domain IQC, i.e., the following inequality holds for all $\Delta_H \in \boldsymbol{\Delta}(m, L)$ and all $y \in \ell_2^p$:

$$\left\langle \begin{bmatrix} y \\ \Delta_H(y) \end{bmatrix}, W^\top \begin{bmatrix} 0 & M^\top \\ M & 0 \end{bmatrix} W \begin{bmatrix} y \\ \Delta_H(y) \end{bmatrix} \right\rangle \geq 0. \tag{3.34}$$

A corresponding set of frequency domain IQCs can be obtained under additional assumptions on the operator M. We elaborate on that later and first concentrate on deriving an analog of the inequality (3.34) for the transformed uncertainty $\Delta_{H,\rho}$. To this end, in the next Lemma we first present a general result building the basis for deriving IQCs for transformed uncertainties from standard ones. A proof is provided in Appendix B.2.5.

Lemma 7. Let $\boldsymbol{\Delta} \subset \{\Delta \in \mathcal{L}(\ell_2^p, \ell_2^p) \mid \Delta(\ell_{2,\rho}^p) \subset \ell_{2,\rho}^p\}$ be some set of operators. Suppose we have given $\phi : \ell_f^{2p} \to \ell_e^{2p}$ such that $\tilde{\phi} = \rho_+ \circ \phi \circ \rho_+$ is bounded on ℓ_2^{2p} and

$$\left\langle \begin{bmatrix} y \\ \Delta(y) \end{bmatrix}, O_T \phi O_T \begin{bmatrix} y \\ \Delta(y) \end{bmatrix} \right\rangle \geq 0 \tag{3.35}$$

for all $y \in \ell_2^p$, $\Delta \in \boldsymbol{\Delta}$, $T \in \mathbb{N}$. Then

$$\left\langle \begin{bmatrix} y \\ \Delta_\rho(y) \end{bmatrix}, \tilde{\phi} \begin{bmatrix} y \\ \Delta_\rho(y) \end{bmatrix} \right\rangle \geq 0 \tag{3.36}$$

holds for all $y \in \ell_2^p$ and any $\Delta_\rho = \rho_- \circ \Delta \circ \rho_+$, $\Delta \in \boldsymbol{\Delta}$. •

Note that (3.35) corresponds to a so-called hard IQC. If $\phi : \ell_2^{2p} \to \ell_2^{2p}$ is bounded on ℓ_2^{2p} and $\tilde{\phi}$ is bounded, then we can replace $O_T \phi O_T$ by ϕ itself in (3.35), i.e.

$$\left\langle \begin{bmatrix} y \\ \Delta(y) \end{bmatrix}, \phi \begin{bmatrix} y \\ \Delta(y) \end{bmatrix} \right\rangle \geq 0 \tag{3.37}$$

implies (3.36).

We next employ Lemma 7 to derive IQCs for the set of transformed uncertainties $\boldsymbol{\Delta}_\rho(m, L)$. In the sequel we limit ourselves to Toeplitz type operators $\bar{M}$. To this end, we introduce the shorthand notation $\mathrm{Toep}(R) : \ell_f^p \to \ell_e^p$ defined as

$$\mathrm{Toep}(R) := \begin{bmatrix} R_0 & \dots & R_{\ell_+} & 0 & \dots & & & \\ R_{-1} & R_0 & \dots & R_{\ell_+} & 0 & \dots & & \\ R_{-2} & R_{-1} & R_0 & \dots & R_{\ell_+} & 0 & \dots & \\ \vdots & \ddots & \ddots & \ddots & \dots & \ddots & \ddots & \end{bmatrix}, \tag{3.38}$$

where $R = \begin{bmatrix} R_{-\ell_-} & R_{-\ell_-+1} & \dots & R_{\ell_+} \end{bmatrix} \in \mathbb{R}^{p \times p(\ell_- + 1 + \ell_+)}$, $R_k \in \mathbb{R}^{p \times p}$, $k = -\ell_-, -\ell_- + 1, \dots, \ell_+$. Note that $\mathrm{Toep}(R)$ is a causal operator if and only if $\ell_+ = 0$. A combination of Lemma 6 with Lemma 7 then yields the following result, a proof can be found in Appendix B.2.6.

Lemma 8. Let $L \geq m > 0$ and $\ell_-, \ell_+ \in \mathbb{N}$ be given. Let $\rho \in (0,1]$. Let $\mathbf{\Delta}_\rho(m, L)$ be defined according to (3.28) and let $M_\rho : \ell_2^p \to \ell_2^p$ be the bounded linear operator with

$$M_\rho = \text{Toep}(\begin{bmatrix} M_{-\ell_-} & M_{-\ell_-+1} & \dots & M_{\ell_+} \end{bmatrix}) \tag{3.39}$$

where $M_i \in \mathbb{R}^{p \times p}$. Then, for all $y \in \ell_2^p$, the inequality (3.34) holds for all $\Delta_H \in \mathbf{\Delta}_\rho(m, L)$ if $M = M_\rho$, $M_i = m_i I_p$ and

$$\text{Toep}(\begin{bmatrix} \rho^{-\ell_-} m_{-\ell_-} & \dots & m_0 & \dots & \rho^{\ell_+} m_{\ell_+} \end{bmatrix}) \tag{3.40}$$

is doubly hyperdominant, i.e.,

$$\sum_{i=-\ell_-}^{\ell_+} m_i \rho^{-i} \geq 0, \qquad \sum_{i=-\ell_-}^{\ell_+} m_i \rho^{i} \geq 0, \tag{3.41}$$

and $m_i \leq 0$ for all $i \in \{-\ell_-, \dots, \ell_+\}, i \neq 0$. •

The Toeplitz operator M_ρ from (3.39) is fully described by the (finite) set of matrices M_i, $i \in \{-\ell_-, -\ell_- + 1, \dots, \ell_+\}$, and, in view of the previous result, we then define the set of admissible matrices as

$$\begin{aligned} \mathbb{M}(\rho, \ell_+, \ell_-, p) = \{ \begin{bmatrix} M_{-\ell_-} \; M_{-\ell_-+1} \cdots M_{\ell_+} \end{bmatrix} \mid \; & M_i = m_i I_p, \; m_i \leq 0 \text{ for all } i \neq 0 \\ & (\textstyle\sum_{i=-\ell_-}^{\ell_+} M_i \rho^{-i})\mathbf{1} \geq 0, \\ & \mathbf{1}^\top (\textstyle\sum_{i=-\ell_-}^{\ell_+} M_i \rho^{i}) \geq 0 \}. \end{aligned} \tag{3.42}$$

We note that the above constraints can be formulated in a simpler fashion in terms of the parameters m_i; we deliberately keep this somewhat cumbersome notation to emphasize the similarities if compared to the structured case discussed in Section 3.4, cf. (3.88), (3.89).

As already indicated before, (3.34) can be interpreted as a time domain IQC. For the sake of embedding the results in the standard setup, we next discuss how to obtain a frequency domain representation. First observe that, for any $y \in \ell_2^p$, we have

$$\begin{aligned} \widehat{My}(z) &= \sum_{t=0}^{\infty} \Big(\sum_{\substack{j=-\ell_+ \\ t-j \geq 0}}^{\ell_-} M_{-j} y_{t-j} \Big) z^{-t} \quad = \sum_{j=-\ell_+}^{\ell_-} M_{-j} \Big(\sum_{\substack{t=0 \\ t-j \geq 0}}^{\infty} y_{t-j} z^{-t} \Big) \\ &= \Big(\sum_{j=-\ell_+}^{\ell_-} M_{-j} z^{-j} \Big) \Big(\sum_{t=0}^{\infty} y_t z^{-t} \Big) = E_M(z) \hat{y}(z), \end{aligned} \tag{3.43}$$

where

$$E_M(z) = \sum_{j=-\ell_+}^{\ell_-} M_{-j} z^{-j}. \tag{3.44}$$

In the same vein, $\widehat{M^\top y}(z) = E_M(z)^* \widehat{y}(z)$. By Parseval's Theorem, we can then formulate (3.34) equally well in the frequency domain as

$$\int_0^{2\pi} \begin{bmatrix} \star \end{bmatrix}^* \widehat{W}^\top \begin{bmatrix} 0 & E_M(e^{i\omega})^* \\ \star & 0 \end{bmatrix} \widehat{W} \begin{bmatrix} \widehat{y}(e^{i\omega}) \\ \widehat{\Delta_H(y)}(e^{i\omega}) \end{bmatrix} d\omega \geq 0, \tag{3.45}$$

where $\widehat{W} \in \mathbb{R}^{2p \times 2p}$ given by

$$\widehat{W} = \begin{bmatrix} (L-m)I & -I \\ 0 & I \end{bmatrix} \tag{3.46}$$

is the z-transform corresponding to the operator defined by (3.33). The conditions on the operator M then translate to conditions on the corresponding transfer matrix E_M. More precisely, the corresponding class of Zames-Falb multipliers is given as

$$\begin{aligned} \mathbf{\Pi}^p_{\Delta,\rho}(m,L) = \{\Pi = \widehat{W}^\top \begin{bmatrix} 0 & E^* \\ E & 0 \end{bmatrix} \widehat{W} \mid E(z) = \textstyle\sum_{j=-\ell_-}^{\ell_+} M_j z^j, \ell_+, \ell_- \in \mathbb{N}, \\ \begin{bmatrix} M_{-\ell_-} & \dots & M_{\ell_+} \end{bmatrix} \in \mathbb{M}(\rho, \ell_+, \ell_-, p)\}, \end{aligned} \tag{3.47}$$

where $\mathbb{M}$ is defined in (3.42). Summing up the previous discussions, we then have the following result concluding Step (S2) in the procedure.

Theorem 6. Let $L \geq m > 0$ and let $\mathbf{\Delta}_\rho(m,L)$ be defined as in (3.28). Then, for each $\rho \in (0,1]$, Δ_ρ satisfies the IQC defined by Π for each $\Delta_\rho \in \mathbf{\Delta}_\rho(m,L)$ and each $\Pi \in \mathbf{\Pi}^p_{\Delta,\rho}(m,L)$ as defined in (3.47). •

Remark 9. Theorem 6 is a generalization of the IQCs presented in Boczar et al. (2015), Lessard et al. (2016) also allowing for anticausal multipliers Π, i.e., anticausal transfer matrices E in (3.47). The class of multipliers from Boczar et al. (2015), Lessard et al. (2016) is then obtained by letting $\ell_+ = 0$. We note that an extension to anticausal multipliers has also been proposed in Freeman (2018) using a slightly different approach. However, the conditions derived in Freeman (2018) are more restrictive, thus leading to a smaller class of multipliers compared to the one proposed here, see Appendix C.2.1 for a discussion. As another advantage, the presented approach is based on standard results and allows for easy extensions such as, e.g., incorporating structural properties of the objective function, see Section 3.4. •

3.2.6 Multiplier parametrization

Having a suitable set of IQCs available as it is provided by Theorem 6, applying Theorem 5 then basically amounts to checking the FDIs (3.29a), (3.29b). In order to derive efficiently implementable conditions in terms of matrix inequalities employing the KYP-Lemma, we require state-space representations of all transfer functions in these FDIs (Step (S3)). This procedure is standard and included here for the sake of completeness; for the multipliers proposed in Freeman (2018) this step is also discussed in detail in Zhang et al. (2019). First note that each transfer matrix E in the definition (3.47) can be decomposed into a causal, an anticausal and a static part, i.e.,

$$E(z) = \sum_{j=-\ell_-}^{-1} M_j z^j + \sum_{j=1}^{\ell_+} M_j z^j + M_0 = E_-(z) + E_+(z) + E_0. \tag{3.48}$$

We then define

$$M_+ = \begin{bmatrix} M_1^\top \; M_2^\top \; \dots \; M_{\ell_+}^\top \end{bmatrix}, \qquad M_- = \begin{bmatrix} M_{-\ell_-} \; M_{-\ell_-+1} \; \cdots \; M_{-1} \end{bmatrix}, \tag{3.49a}$$

$M_+ \in \mathbb{R}^{p \times \ell_+ p}, M_- \in \mathbb{R}^{p \times \ell_- p}$, as well as the vectors of strictly proper and stable basis functions $\psi_+ \in \mathcal{RH}_\infty^{\ell_+ \times 1}, \psi_- \in \mathcal{RH}_\infty^{\ell_- \times 1}$ given by

$$\psi_+(z) = \begin{bmatrix} z^{-1} & z^{-2} & \dots & z^{-\ell_+} \end{bmatrix}^\top, \qquad \psi_-(z) = \begin{bmatrix} z^{-\ell_-} & z^{-\ell_-+1} & \dots & z^{-1} \end{bmatrix}^\top. \tag{3.50a}$$

Hence, $E_+(z)^* = M_+(\psi_+(z) \otimes I_p)$ as well as $E_-(z) = M_-(\psi_-(z) \otimes I_p)$. Note that $\ell_+ = 0$ and $\ell_- = 0$ correspond to the case where the transfer function has no causal or no anticausal part. Note further that, in general, also other types of basis functions can be employed which might result in smaller values of ℓ_+, ℓ_- yielding similar outcomes which is desired in numerical implementations. The standard trick to deal with the anticausal part is then to factorize the multiplier as

$$\widehat{W}^\top \begin{bmatrix} 0 & E(z)^* \\ E(z) & 0 \end{bmatrix} \widehat{W} = \psi_\Delta(z)^* M_\Delta \psi_\Delta(z), \tag{3.51}$$

where

$$M_\Delta = \left[\begin{array}{cc|cc|cc} 0 & M_0^\top & 0 & 0 & 0 & 0 \\ M_0 & 0 & 0 & 0 & 0 & 0 \\ \hline 0 & 0 & 0 & M_-^\top & 0 & 0 \\ 0 & 0 & M_- & 0 & 0 & 0 \\ \hline 0 & 0 & 0 & 0 & 0 & M_+ \\ 0 & 0 & 0 & 0 & M_+^\top & 0 \end{array}\right], \tag{3.52}$$

$$\psi_\Delta(z) = \left[\begin{array}{cc} I_p & 0 \\ 0 & I_p \\ \hline \psi_-(z) \otimes I_p & 0 \\ 0 & I_p \\ \hline I_p & 0 \\ 0 & \psi_+(z) \otimes I_p \end{array}\right] \widehat{W}. \tag{3.53}$$

Note that ψ_Δ is proper and stable and, in the following, we denote by $(A_\Delta, B_\Delta, C_\Delta, D_\Delta)$ its state-space realization when using the state-space realizations of ψ_+, ψ_- provided in Appendix C.2.2.

3.3 Robust Optimization Algorithms

3.3.1 Algorithm Analysis

3.3.1.1 Convergence Rate Analysis

We begin with deriving convex conditions in terms of linear matrix inequalities (LMIs) for determining convergence rate bounds for a given algorithm in the form (3.2). In a nutshell, this amount to reformulating the FDI (3.29a) using the KYP-Lemma (see Lemma 16). Let $G_{\tilde{y}\tilde{w}}$ be defined according to (3.22). We then define the following state-space realization

$$\psi_c = \psi_\Delta \begin{bmatrix} G_{\tilde{y}\tilde{w}} \\ I_p \end{bmatrix} \sim (A_c(\rho), B_c, C_c(\rho), D_c), \tag{3.54}$$

where $A_c(\rho) \in \mathbb{R}^{n_c \times n_c}$, $B_c \in \mathbb{R}^{n_c \times p_c}$, $C_c(\rho) \in \mathbb{R}^{q_c \times n_c}$, $D_c \in \mathbb{R}^{q_c \times p_c}$ with $n_c = p(n + \ell_- + \ell_+)$, $p_c = p$, $q_c = p(4 + \ell_- + \ell_+)$. A particular state-space realization is given in Appendix C.2.2 and in the sequel we assume $A_c(\rho), B_c, C_c(\rho), D_c$ to be given in that form. The following result then virtually follows immediately from Theorem 5 employing the class of multipliers introduced in Theorem 6. A proof is given in Appendix B.2.7.

Theorem 7. Consider (3.2). Suppose $A \in \mathbb{R}^{np \times np}$, $B \in \mathbb{R}^{np \times p}$, $C \in \mathbb{R}^{p \times np}$, $D \in \mathbb{R}^{np \times p}$, $D^{\dagger} \in \mathbb{R}^{np \times p}$ are given and fulfill (3.4). Let $L \geq m > 0$ be given. Fix $\rho \in (0,1)$ and assume that $A + mBC$ has all eigenvalues in the open disk of radius ρ. Let some $\ell_+, \ell_- \in \mathbb{N}$ be given and let M_Δ, $\mathbb{M}$ be defined according to (3.52), (3.42). If there exist $P = P^\top \in \mathbb{R}^{n_c \times n_c}$, $M_+ \in \mathbb{R}^{p \times \ell_+ p}$, $M_- \in \mathbb{R}^{p \times \ell_- p}$, $M_0 \in \mathbb{R}^{p \times p}$ such that

$$\begin{bmatrix} \\ \star \\ \\ \end{bmatrix}^\top \begin{bmatrix} P & 0 & 0 \\ 0 & -P & 0 \\ 0 & 0 & M_\Delta \end{bmatrix} \begin{bmatrix} A_c(\rho) & B_c \\ I & 0 \\ C_c(\rho) & D_c \end{bmatrix} \prec 0, \tag{3.55a}$$

$$\begin{bmatrix} M_- & M_0 & M_+ \end{bmatrix} \in \mathbb{M}(\rho, \ell_+, \ell_-, p), \tag{3.55b}$$

then the origin is robustly exponentially stable against $\mathbf{\Delta}(m, L)$ for (3.15), thus, for all $H \in \mathcal{S}_{m,L}$, the output z_k of (3.2) converges with rate ρ to the unique minimizer $z^\star$ of (3.1). •

For a given optimization algorithm described by matrices A, B, C, D, Theorem 7 allows checking whether the algorithm converges exponentially with rate less or equal than some given rate ρ by trying to find a feasible solution to the LMIs (3.55). In convergence rate analysis, however, we are typically interested in finding the smallest ρ such that (3.55) is feasible. If ρ is kept as a free variable, the inequalities are no longer linear in the unknown parameters. Still, if (3.55) is feasible for some $\rho \in (0,1)$, then it also feasible for all $\bar{\rho} \in [\rho, 1]$. Therefore, it is convenient to do a bisection search over ρ to find the optimal convergence rate.

Many optimization algorithms admit the structure $A = \overline{A} \otimes I_p$, $B = \overline{B} \otimes I_p$, $C = \bar{C} \otimes I_p$, $D = \overline{D} \otimes I_p$. In that case we can also find a state-space realization $(A_c(\rho), B_c, C_c(\rho), D_c)$ such that $A_c = \overline{A_c}(\rho) \otimes I_p$, $B_c = \overline{B_c} \otimes I_p$, $C_c = \overline{C_c}(\rho) \otimes I_p$, $D_c = \overline{D_c} \otimes I_p$. Note that also

$M_\Delta(M_+, M_-, M_0) = M_\Delta(m_+, m_-, m_0) \otimes I_p$. In this situation we can take $P = \overline{P} \otimes I_p$ without loss of generality, i.e., instead of (3.55a) we can solve the following LMI

$$\begin{bmatrix} \\ \star \\ \\ \end{bmatrix}^\top \begin{bmatrix} \overline{P} & 0 & 0 \\ 0 & -\overline{P} & 0 \\ 0 & 0 & M_\Delta \end{bmatrix} \begin{bmatrix} \overline{A_c}(\rho) & \overline{B_c} \\ I & 0 \\ \overline{C_c}(\rho) & \overline{D_c} \end{bmatrix} \prec 0, \tag{3.56a}$$

$$\begin{bmatrix} m_- & m_0 & m_+ \end{bmatrix} \in \mathbb{M}(\rho, \ell_+, \ell_-, 1), \tag{3.56b}$$

where now the dimension is reduced by a factor of p compared to (3.55a). This dimensionality reduction is lossless in the sense that (3.55) are feasible if and only if (3.56) are feasible. We provide a proof of this statement in Appendix B.2.8.

3.3.1.2 Performance Analysis

We next derive similar convex conditions for analyzing the performance of a given algorithm by analogously reformulating the FDI (3.29b). While the methodology applies to a much more general class of performance specifications such as H_∞-performance, we concentrate on an adapted H_2-performance measure here and show that this can be related to noise rejection properties of the optimization algorithm.

We first recall that the H_2-norm of a stable linear time-invariant system with strictly proper transfer matrix G is defined as

$$\|G\|_2^2 = \tfrac{1}{2\pi}\mathrm{tr}\Big(\int_0^{2\pi} G(e^{\mathrm{i}\omega})^* G(e^{\mathrm{i}\omega})\mathrm{d}\omega\Big). \tag{3.57}$$

This definition of the H_2-norm is mainly motivated by the following two interpretations: first, the H_2-norm is a measure for the energy of the system's impulse response; second, it can as well be interpreted as the asymptotic variance of the output when the system is driven by white noise. However, these interpretations do not directly carry over to nonlinear or time-varying systems (Paganini & Feron, 2000). Different extensions have been proposed, which are rather based on the actual desired performance measure instead of a certain system norm. In the following we build upon a stochastic interpretation; this is motivated by the application problem at hand as we will discuss in more detail in Section 3.3.1.3. We next give a precise definition of the proposed performance measure.

Definition 12 (Asymptotic H_2-performance). Consider the state-space representation (3.17) of the feedback interconnection (3.5) and let $w_\mathrm{p} = (W_t)_{t\in\mathbb{N}}$ be a discrete-time white noise process. Let $(y_{\mathrm{p},t})_{t\in\mathbb{N}}$ denote the corresponding response of (3.17) with initial condition $x_0 = 0$. We then say that the feedback interconnection (3.5) *achieves a robust H_2-performance level of $\gamma > 0$ against* $\mathbf{\Delta}$ if (3.5) is robustly stable against $\mathbf{\Delta}$ and if, for all $\Delta \in \mathbf{\Delta}$, the performance channel fulfills $\|G_{y_\mathrm{p}w_\mathrm{p}}\|_{2,\mathrm{av}} \leq \gamma$, where

$$\|G_{y_\mathrm{p}w_\mathrm{p}}\|_{2,\mathrm{av}} = \limsup_{t_{\max}\to\infty} \sqrt{\tfrac{1}{t_{\max}} \sum_{t=0}^{t_{\max}} \mathrm{E}(y_{\mathrm{p},t}^\top y_{\mathrm{p},t})} \tag{3.58}$$

and $\mathrm{E}(Y)$ is the expected value of a random variable Y. •

A similar measure has been proposed in Paganini and Feron (2000). The motivation for this definition is that a small value of this asymptotic H_2-measure can be related to the noise rejection properties of the system under consideration. If the feedback interconnection (3.5) is linear time invariant, then this definition directly carries over to a standard H_2-norm constraint (3.57).

We next discuss how the so-defined robust H_2-performance can be analyzed within the IQC framework. We consider performance multipliers that admit a parametrization of the following form

$$\boldsymbol{\Pi}_{\mathrm{p}} = \left\{\psi_{\mathrm{p}}^* M_{\mathrm{p}} \psi_{\mathrm{p}} : M_{\mathrm{p}} = \begin{bmatrix} M_{\mathrm{p},11} & M_{\mathrm{p},12} \\ M_{\mathrm{p},12}^{\top} & M_{\mathrm{p},22} \end{bmatrix} \in \mathbf{M}_{\mathrm{p}}, M_{\mathrm{p},22} \succeq 0\right\}, \tag{3.59}$$

where $\mathbf{M}_{\mathrm{p}} \subset \mathbb{R}^{q\times q}$ is some convex set and $\psi_{\mathrm{p}} \in \mathcal{RH}_{\infty}^{q\times(n_{w_{\mathrm{p}}}+n_{y_{\mathrm{p}}})}$, $q \in \mathbb{N}_{>0}$, is a proper and stable transfer function. We emphasize that the most relevant performance measures fall into that class, e.g., H_∞- or the adapted H_2-performance measure as considered here. Using the factorization of the Zames-Falb multipliers from (3.51), the FDI for performance (3.29b) is given by

$$\psi_{\mathrm{c}}^* \left[\begin{array}{c|c} M_\Delta & 0 \\ \hline 0 & M_{\mathrm{p}} \end{array}\right] \psi_{\mathrm{c}} \overset{\mathbb{T}}{\prec} 0, \tag{3.60}$$

where

$$\psi_{\mathrm{c}} = \left[\begin{array}{c|c} \psi_\Delta & 0 \\ \hline 0 & \psi_{\mathrm{p}} \end{array}\right] \left[\begin{array}{c|c} G_{yw} & G_{yw_{\mathrm{p}}} \\ I_p & 0 \\ \hline 0 & I_{n_{w_{\mathrm{p}}}} \\ G_{y_{\mathrm{p}}w} & G_{y_{\mathrm{p}}w_{\mathrm{p}}} \end{array}\right]. \tag{3.61}$$

Let a state-space realization be given by

$$\psi_{\mathrm{c}} \sim \left(\begin{array}{c|c} \boldsymbol{A}_{\mathrm{c}} & \begin{bmatrix} \boldsymbol{B}_{\mathrm{c},1} & \boldsymbol{B}_{\mathrm{c},2} \end{bmatrix} \\ \begin{bmatrix} \boldsymbol{C}_{\mathrm{c},1} \\ \boldsymbol{C}_{\mathrm{c},2} \end{bmatrix} & \begin{bmatrix} \boldsymbol{D}_{\mathrm{c},11} & \boldsymbol{D}_{\mathrm{c},12} \\ \boldsymbol{D}_{\mathrm{c},21} & \boldsymbol{D}_{\mathrm{c},22} \end{bmatrix} \end{array}\right). \tag{3.62}$$

We emphasize that such a state-space realization can easily be obtained using state-space realizations of the single transfer functions appearing in (3.61). For implementation purpose, this is particularly easy making use of numerical tools such as the control systems toolbox in MATLAB. For completeness, we give an explicit state-space realization in Appendix C.2.2.

In the particular case of the adapted H_2-performace measure, the performance multiplier is defined with

$$\psi_{\mathrm{p}}(z) = \begin{bmatrix} 0_{n_{y_{\mathrm{p}}}\times n_{w_{\mathrm{p}}}} & I_{n_{y_{\mathrm{p}}}} \end{bmatrix}, \quad \mathbf{M}_{\mathrm{p}} = \left\{M = I_{n_{y_{\mathrm{p}}}}\right\}, \tag{3.63}$$

and we denote the resulting state-space realization by (3.62). Note that $\mathbf{M}_\mathrm{p}$ is a singleton here and the H_2-performance level is set by the aforementioned additional condition resulting in (3.64b). We then have the following result for H_2-performance; a proof is given in Appendix B.2.9.

Theorem 8 (Robust H_2-performance). Consider (3.15) and suppose $D_\mathrm{p} = 0$. Suppose $A \in \mathbb{R}^{np \times np}$, $B \in \mathbb{R}^{np \times p}$, $C \in \mathbb{R}^{p \times np}$, $D \in \mathbb{R}^{p \times np}$, $D^\dagger \in \mathbb{R}^{np \times p}$ are given and fulfill (3.4). Let $L \geq m > 0$ be given. Assume that $A + mBC$ has all eigenvalues in the open unit disk. Let some $\ell_+, \ell_- \in \mathbb{N}$ be given and let M_Δ be defined according to (3.52). If there exist $P_\mathrm{p} = P_\mathrm{p}^\top \in \mathbb{R}^{n_\mathrm{c} \times n_\mathrm{c}}$, $M_+ \in \mathbb{R}^{p \times p\ell_+}$, $M_- \in \mathbb{R}^{p \times p\ell_-}$, $M_0 \in \mathbb{R}^{p \times p}$, $\gamma > 0$, such that

$$\begin{bmatrix} \\ \star \\ \\ \end{bmatrix}^\top \left[\begin{array}{cc|cc} P_\mathrm{p} & 0 & 0 & 0 \\ 0 & -P_\mathrm{p} & 0 & 0 \\ \hline 0 & 0 & M_\Delta & 0 \\ 0 & 0 & 0 & I_{n_{y\mathrm{p}}} \end{array}\right] \left[\begin{array}{cc} \boldsymbol{A}_\mathrm{c} & \boldsymbol{B}_{\mathrm{c},1} \\ I & 0 \\ \hline \boldsymbol{C}_{\mathrm{c},1} & \boldsymbol{D}_{\mathrm{c},11} \\ \boldsymbol{C}_{\mathrm{c},2} & 0 \end{array}\right] \prec 0 \tag{3.64a}$$

$$\mathrm{tr}(\boldsymbol{B}_{\mathrm{c},2}^\top N N^\top P_\mathrm{p} N N^\top \boldsymbol{B}_{\mathrm{c},2}) \leq \gamma^2 \tag{3.64b}$$

$$P_\mathrm{p} \succ 0 \tag{3.64c}$$

$$N^\top = \begin{bmatrix} 0_{np \times (\ell_- + \ell_+)p} & I_{np} \end{bmatrix} \tag{3.64d}$$

$$\begin{bmatrix} M_- & M_0 & M_+ \end{bmatrix} \in \mathbb{M}(1, \ell_+, \ell_-, p), \tag{3.64e}$$

then the interconnection (3.15) is robustly stable against $\boldsymbol{\Delta}(m, L)$ and it achieves a robust H_2-performance level of γ against $\boldsymbol{\Delta}(m, L)$. •

Remark 10. It is well-known that under the positive definiteness condition (3.64c) the trace condition (3.64b) holds if and only if there exists a matrix Z of suitable dimensions such that

$$\begin{bmatrix} N^\top P_\mathrm{p} N & N^\top P_\mathrm{p} N N^\top \boldsymbol{B}_{\mathrm{c},2} \\ \star & Z \end{bmatrix} \succ 0, \quad \mathrm{tr}(Z) < \gamma^2. \tag{3.65}$$

To see this, note that, by taking Schur complements, the above conditions hold if and only if

$$N^\top P_\mathrm{p} N \succ 0, \quad Z \succ \boldsymbol{B}_{\mathrm{c},2}^\top N N^\top P_\mathrm{p} N N^\top \boldsymbol{B}_{\mathrm{c},2} \tag{3.66}$$

and $\mathrm{tr}(Z) < \gamma^2$. If the latter inequalities are feasible, this implies that there exists an $\varepsilon > 0$ such that $Z = \boldsymbol{B}_{\mathrm{c},2}^\top N N^\top P_\mathrm{p} N N^\top \boldsymbol{B}_{\mathrm{c},2} + \varepsilon I$ is a valid solution. Then this implies that (3.64b) holds. •

We additionally note that, as already discussed for the convergence rate test from Theorem 7, an analogous lossless dimensionality reduction is possible for H_2-performance analysis for equally structured optimization algorithms whenever the matrices defining the performance channel admit the same structure, i.e., $B_\mathrm{p} = \overline{B_\mathrm{p}} \otimes I_p$, $C_\mathrm{p} = \overline{C_\mathrm{p}} \otimes I_p$, $D_\mathrm{p} = \overline{D_\mathrm{p}} \otimes I_p$. This structure then also transfers to the matrices $\boldsymbol{A}_\mathrm{c}, \boldsymbol{B}_\mathrm{c}, \boldsymbol{C}_\mathrm{c}, \boldsymbol{D}_\mathrm{c}$.

	Parameters			Convergence Rate	Reference
	ν_1	ν_2	ν_3	ρ	
GD	$\frac{2}{m+L}$	0	0	$\frac{\kappa-1}{\kappa+1}$	Polyak (1987)
NM	$\frac{1}{L}$	$\frac{\sqrt{L}-\sqrt{m}}{\sqrt{L}+\sqrt{m}}$	ν_2	$\sqrt{1-\frac{1}{\sqrt{\kappa}}}$	Nesterov (2004)
NM$_{\text{mod}}$	$\frac{1}{L}$	$\frac{2\kappa-\sqrt{2\kappa-1}-1}{2(\kappa+\sqrt{\kappa-1})}$	ν_2	$\sqrt{1-\frac{\sqrt{2\kappa-1}}{\kappa}}$	Safavi et al. (2018)
TMM	$\frac{1+\rho}{L}$	$\frac{\rho^2}{2-\rho}$	$\frac{\rho^2}{(1+\rho)(2-\rho)}$	$1-\frac{1}{\sqrt{\kappa}}$	Van Scoy et al. (2018)
HB	$(\frac{2}{\sqrt{L}+\sqrt{m}})^2$	$(\frac{\sqrt{L}-\sqrt{m}}{\sqrt{L}+\sqrt{m}})^2$	0	-	Polyak (1987)

Table 3.1. Parameters of several popular algorithms of the form (3.67) and the corresponding known upper bounds on the convergence rate in dependency of the condition ratio $\kappa = \frac{L}{m}$. Note that in the case of the Gradient Descent algorithm the parameter choice reduces (3.67) to a first order algorithm. NM$_{\text{mod}}$ is a slightly modified version of Nesterov's Method for which an improved convergence rate bound is proven in Safavi et al. (2018). While an explicit convergence rate can be given for the Heavy Ball Method in case of quadratic objective functions, it is known not to be globally convergent for general $H \in \mathcal{S}_{m,L}$, see Lessard et al. (2016) for a counterexample.

3.3.1.3 Numerical results

In the following we use the presented results to numerically analyze existing optimization algorithms in terms of their convergence rate as well as their properties in the presence of additive gradient noise. Here, we consider the Gradient Descent algorithm (GD), Nesterov's Method (NM) with constant step size (Nesterov, 2004), the Triple Momentum Method (TMM) (Van Scoy et al., 2018) as well as the Heavy Ball Method (HB) (Polyak, 1987). In all cases, the matrices A, B, C, D in (3.2) take the form

$$A = \begin{bmatrix} 1+\nu_2 & -\nu_2 \\ 1 & 0 \end{bmatrix} \otimes I_p, \qquad B = \begin{bmatrix} -\nu_1 \\ 0 \end{bmatrix} \otimes I_p, \tag{3.67a}$$

$$C = \begin{bmatrix} 1+\nu_3 & -\nu_3 \end{bmatrix} \otimes I_p, \qquad D = \begin{bmatrix} 1 & 0 \end{bmatrix} \otimes I_p, \tag{3.67b}$$

where the scalar parameters ν_1, ν_2, ν_3 are as given in Table 3.1. Note that with $D^\dagger = \begin{bmatrix} 1 & 1 \end{bmatrix}^\top \otimes I_p$, the conditions (3.4) are fulfilled. Due to the Kronecker structure and by our previous discussion, we set $p = 1$ without loss of generality. The following as well as all other numerical results in this chapter were obtained using MATLAB together with YALMIP (Löfberg, 2004).

Convergence rate analysis For convergence rate analysis, we use Theorem 7 together with a bisection search over ρ to determine upper bounds on the convergence rates for different condition ratios $\kappa = L/m$, see Figure 3.6. For Nesterov's Method, the Gradient Descent algorithm and the Triple Momentum Method, we reproduce the known convergence rate bounds. The numerical results suggest that the improved bound shown in Safavi et al. (2018) for a modified version of Nesterov's method (see Table 3.1) also hold for the standard parameter choice. For the Heavy Ball Method, global convergence can be guaranteed for small condition numbers; for larger condition numbers the LMIs in Theorem 7 turn out to be infeasible, even for $\rho = 1$. These results are in concordance with Lessard et al. (2016). For Nesterov's Method, Gradient Descent and the Triple Momentum Method, the addition of anticausal Zames-Falb multipliers does not lead to any improvement of the upper bound on the convergence rate and it is even sufficient to choose $\ell_+ = 1, \ell_- = 0$. In contrast, for the Heavy Ball Method employing anticausal multipliers with $\ell_+ = 1, \ell_- = 5$ leads to slightly better convergence rate estimates. Using higher dimensions for both the causal and the anticausal part does not significantly improve the bounds. Except for the Gradient Descent algorithm, it is still an open question whether the known as well as the numerically determined bounds are tight in the sense that there exists some objective function $H \in \mathcal{S}_{m,L}$ such that the resulting algorithm does not converge faster than presumed by the given bound.

H_2-performance In the following we investigate the properties of the algorithms from Table 3.1 when the gradient is affected by white noise in an additive fashion, i.e., in rough words, in (3.2) we have $\nabla H(Cx_t) + W_t$ instead of $\nabla H(Cx_t)$, where $(W_t)_{t\in\mathbb{N}}$ is a discrete-time white noise process. Such situations occur in several applications, e.g., when the objective function and its gradient are not evaluated by means of numeric calculations but rather by measurements (Romer, Montenbruck, & Allgöwer, 2017). Likewise, in empirical risk minimization as it is utilized, e.g., in the context of learning algorithms (Murphy & Bach, 2012), the objective function is given by an expected value, which, however, cannot be evaluated since the underlying probability distribution is unknown. It is therefore common to use a sample-based approximation of the expected value, the so-called empirical risk. According to the central limit theorem, this approximation differs from the original expected value by additive random noise, see Polyak (1987). Similar situations appear when employing Monte-Carlo methods. The asymptotic H_2-performance as defined in Definition 12 is a measure for the noise attenuation; hence we choose the performance channel as $B_\mathrm{p} = B, C_\mathrm{p} = D, D_\mathrm{p} = 0$ such that the corresponding H_2-performance level is a measure how additive gradient noise affects the resulting optimizer. We then use Theorem 8 to determine upper bounds on the corresponding H_2-performance. The results are depicted on the left-hand side of Figure 3.7. These numerical results suggest that the Gradient Descent algorithm has the best properties in terms of additive noise attenuation for condition ratios larger than approximately 10, followed by Nesterov's Method. The fastest method in terms of convergence rates, the Triple Momentum Method, however, has the worst noise attenuation. We note that these results have to be taken with care since they only provide upper bounds; still, they are qualitatively in accordance with the results from Mohammadi, Razaviyayn, and Jovanović (2018), where a similar performance channel has been analyzed for quadratic optimization problems.

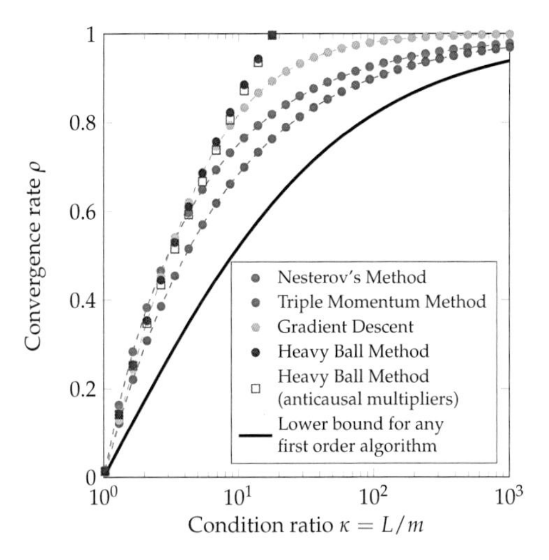

Figure 3.6. Upper bounds on the convergence rates of different algorithms obtained from Theorem 7 for condition ratios between 1.02 and 1000. The corresponding dashed lines indicate the known convergence rate bounds, cf. Table 3.1, where, for Nesterov's Method, the analytical bound for the modified version is depicted. The black line indicates the fundamental lower bound for any first order optimization algorithm for objective functions in $\mathcal{S}_{m,L}$, see Nesterov (2004).

To underpin this statement, we also carried out a sample-based approach to evaluate the H_2-performance. To this end, we randomly generated 10000 functions from the class $\mathcal{S}_{m,L}$ and simulated the four considered optimization algorithms under additional noise. The corresponding lower H_2-performance bound estimates are depicted on the right-hand side of Figure 3.7. While the results are qualitatively similar, still, there is quite a gap quantitatively. We emphasize that this is not a contradiction since, first, Theorem 8 does only provide an upper bound, and second – and probably more important – the sampling of the class $\mathcal{S}_{m,L}$ is not very dense. In fact, we only sample quadratic functions $H \in \mathcal{S}_{m,L}$ as well as functions with a Hessian of the form $\nabla^2 H(z) = c_1 + c_2 \cos(\omega z)$, where $m \leq c_1 + c_2 \leq L$, $\omega > 0$ are chosen randomly.

The previous results suggest that there is a trade-off between convergence speed and robustness towards noise, an observation that has also been made in the literature in different settings, see Aybat et al. (2019); Polyak (1987) or Lessard et al. (2016) in the case of relative deterministic noise. Our numerical results underpin and quantify these findings. This observation becomes even more apparent when designing novel optimization algorithms, see Section 3.3.2.3. Similarly to the convergence rate analysis, anticausal Zames-Falb multipliers do not lead to improved bounds.

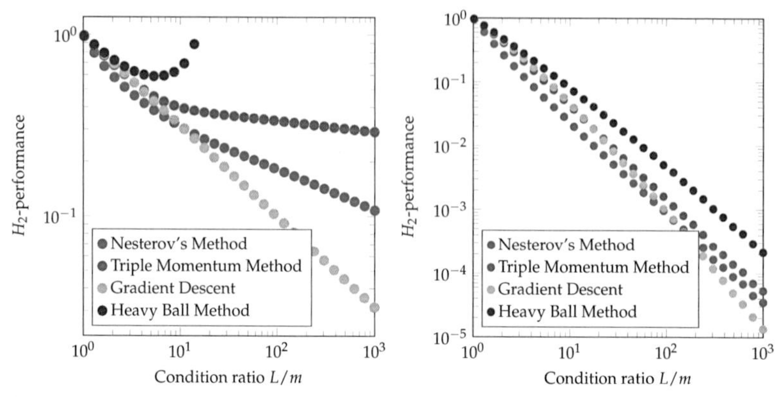

Figure 3.7. Upper bounds on the H_2-performance related to additive gradient noise for different optimization algorithms and condition ratios between 1.02 and 1000 obtained using Theorem 8 (left) and lower bounds from a sample-based approach (right). In all cases, the dimensions of the Zames-Falb multipliers were chosen as $\ell_+ = 0$, $\ell_- = 4$.

3.3.2 Algorithm Synthesis

In the following we consider the problem of designing algorithms that have desired properties specified in terms of convergence rates and H_2-performance bounds. To this end, roughly speaking, we need to combine Theorem 7 and Theorem 8 with the difference that the algorithm parameters A, B, C, D are now to be determined as well. Motivated by our discussions in Section 3.1, we fix C, D and $D^\dagger$ to

$$C = \begin{bmatrix} 1 & 0 & \dots & 0 \end{bmatrix} \otimes I_p, \quad D = C, \quad D^\dagger = C^\top. \tag{3.68}$$

Still, having A, B as additional design variables, the following main difficulties compared to the analysis problem need to be addressed: (1) The conditions (3.4) have to be ensured to hold; (2) nominal exponential stability of the feedback interconnection has to be guaranteed, i.e., $A + mBC$ must have all its eigenvalues in the open disk of radius ρ; and (3) the introduction of new design variables leads to bilinear instead of linear matrix inequalities in (3.55), (3.64), which, in general, cannot be solved efficiently. We emphasize that in the end we are interested in finding LMI conditions, which is the core difficulty of all three problems. As it turns out, (1) can be handled without much effort and, as we show next, (2) can be resolved utilizing the special structure of the problem at hand.

Lemma 9 (Nominal exponential stability). Let $\rho \in (0, 1)$ be given and consider A_c, B_c, C_c, D_c as defined by (3.54), cf. (C.26). Suppose that there exists P partitioned as

$$P = \begin{bmatrix} P_{11} & P_{12} \\ P_{12}^\top & P_{22} \end{bmatrix}, P_{11} \in \mathbb{R}^{p(\ell_- + \ell_- + 1) \times p(\ell_- + \ell_- + 1)}, P_{22} \in \mathbb{R}^{np \times np}, \tag{3.69}$$

such that (3.55a) holds. Then $A + mBC$ has all eigenvalues in the open disk of radius ρ if and only if $P_{22} \succ 0$. •

Proof. Consider the lower right block of (3.55a). The inequality then implies that

$$\rho^{-2}(A + mBC)^\top P_{22}(A + mBC) - P_{22} + \rho^{-2}C_{\psi_1}^\top D_\Delta^\top M_\Delta(M_+, M_-, M_0)D_\Delta C_{\psi_1} \prec 0. \quad (3.70)$$

By straightforward calculations, it is seen that $C_{\psi_1}^\top D_\Delta^\top M_\Delta(M_+, M_-, M_0)D_\Delta C_{\psi_1} = 0$; thus, by standard Lyapunov theory for linear systems, we infer that $A + mBC$ has all eigenvalues in the open disk of radius ρ if and only if $P_{22} \succ 0$. □

Remark 11. The same applies for robust performance, i.e., a similar partitioning of P_p together with an appropriate positive definiteness constraint ensures that (3.64a) implies nominal stability. We omit a precise statement and a proof here since typically in synthesis problems we are interested in both robust exponential stability and robust performance. •

In order to address (3), it is convenient to first reformulate the matrix inequalities by means of Schur complements. As apparent in (3.70), with A, B being design variables in the synthesis case, (3.55a) is a quadratic matrix inequality in the unknowns. The same holds true for the performance inequalities (3.64a). As a first step, we utilize the positive definiteness condition for nominal exponential stability and a similar condition for performance in order to reformulate these quadratic matrix inequalities to bilinear matrix inequalities (BMIs). Note that with (3.64d) and (3.69) we have

$$P_{22} = N^\top P N; \quad (3.71)$$

hence the positive definiteness constraint $P_{22} \succ 0$ allows us to equivalently formulate (3.55a) as (see Appendix C.2.3 for a derivation)

$$\begin{bmatrix} -P_{22} & P_{22}N^\top \begin{bmatrix} A_\mathrm{c}(\rho) & B_\mathrm{c} \end{bmatrix} \\ \star & U(P, M_\Delta) \end{bmatrix} \prec 0, \quad (3.72)$$

where we introduced the shorthand notation

$$U(P, M_\Delta) = \begin{bmatrix} \star \end{bmatrix}^\top \begin{bmatrix} P - NP_{22}N^\top & 0 & 0 \\ 0 & -P & 0 \\ 0 & 0 & M_\Delta \end{bmatrix} \begin{bmatrix} A_\mathrm{c}(\rho) & B_\mathrm{c} \\ I & 0 \\ C_\mathrm{c} & D_\mathrm{c} \end{bmatrix}. \quad (3.73)$$

Note that, by the structure of the state-space representation (C.26), (3.72) is bilinear in the unknowns A, B, P, M_Δ.

A similar procedure applies to the problem with additional performance specifications. More precisely, assuming P_p to be structured in the same manner as P, (3.64a), (3.64c) are equivalently formulated as

$$\begin{bmatrix} -P_{\mathrm{p},22} & P_{\mathrm{p},22}N^\top \begin{bmatrix} \boldsymbol{A}_\mathrm{c} & \boldsymbol{B}_{\mathrm{c},1} \end{bmatrix} \\ \star & U_\mathrm{p}(P_\mathrm{p}, M_\Delta) \end{bmatrix} \prec 0, \quad (3.74)$$

where

$$U_\mathrm{p}(P_\mathrm{p}, M_\Delta) = \begin{bmatrix} \star \end{bmatrix}^\top \left[\begin{array}{cc|cc} P_\mathrm{p} - NP_{\mathrm{p},22}N^\top & 0 & 0 & 0 \\ 0 & -P_\mathrm{p} & 0 & 0 \\ \hline 0 & 0 & M_\Delta & 0 \\ 0 & 0 & 0 & I_{n_{y\mathrm{p}}} \end{array}\right] \left[\begin{array}{cc} \boldsymbol{A}_\mathrm{c} & \boldsymbol{B}_{\mathrm{c},1} \\ I & 0 \\ \hline \boldsymbol{C}_{\mathrm{c},1} & \boldsymbol{D}_{\mathrm{c},11} \\ \boldsymbol{C}_{\mathrm{c},2} & 0 \end{array}\right]. \tag{3.75}$$

In the following we consider a performance channel defined as in the example in Section 3.3.1.3, i.e., $B_\mathrm{p} = B$, $C_\mathrm{p} = C$, $D_\mathrm{p} = 0$. Note that, by this choice, B_p is as well a design variable, while $C_\mathrm{p}, D_\mathrm{p}$ are fixed, and (3.74) is bilinear in the unknowns A, B, P, M_Δ. Additionally, keeping in mind Remark 10 and employing the specific state-space realization from Appendix C.2.2, (3.64b) can be reformulated as

$$\begin{bmatrix} P_{\mathrm{p},22} & P_{\mathrm{p},22}N^\top \boldsymbol{B}_{\mathrm{c},2} \\ \star & Z \end{bmatrix} \succ 0, \quad \mathrm{tr}(Z) < \gamma^2. \tag{3.76}$$

Still, for the synthesis problem the former inequalities are bilinear. We next discuss two approaches to cope with that fact. The first one relies on rendering the conditions linear by imposing certain restrictions on the design variables and employing a suitable variable transformation; the second one is an often used rather hands-on heuristic based on alternately solving LMIs to obtain feasible solutions of the BMIs.

3.3.2.1 Convex solution

We next discuss how to obtain LMI conditions from the BMIs (3.72), (3.74), (3.76). Variable transformations have successfully been employed in many situations to render the inequalities linear in the new variables. However, since $A_\mathrm{c}(\rho), B_\mathrm{c}, \boldsymbol{A}_\mathrm{c}, \boldsymbol{B}_\mathrm{c}$ are structured matrices here, the standard variable transformations do not directly apply. To make this more vivid, consider (3.72). With A, B, P being design variables, a typical approach is to define $Q_A = P_{22}A$, $Q_B = P_{22}B$. However, U as defined in (3.73) also contains nonlinear terms of the form $P_{12}A = P_{12}P_{22}^{-1}Q_A$, $P_{12}B = P_{12}P_{22}^{-1}Q_B$, where P_{12} is the right upper block of P, see (3.69). These terms cannot be handled by that approach; a simple remedy is to let $P_{12} = 0$, i.e., P is block diagonal, which is the idea behind the following Theorem.

Theorem 9. Let $L \geq m > 0$ as well as $n \in \mathbb{N}_{>0}$, $p \in \mathbb{N}_{>0}$ be given. Let $C \in \mathbb{R}^{p\times np}$, $D \in \mathbb{R}^{np\times p}$, $D^\dagger \in \mathbb{R}^{p\times np}$ be defined as in (3.68). Fix $\rho \in (0,1)$. Let some $\ell_-, \ell_+ \in \mathbb{N}$ be given and let M_Δ, $\mathbb{M}$ be defined according to (3.52), (3.42). Suppose there exist $P_{11} = P_{11}^\top \in \mathbb{R}^{p(\ell_-+\ell_+)\times p(\ell_-+\ell_+)}$, $P_{22} = P_{22}^\top \in \mathbb{R}^{np\times np}$, $Q_A \in \mathbb{R}^{np\times np}$, $Q_B \in \mathbb{R}^{np\times p}$, $M_+ \in \mathbb{R}^{p\times\ell_- p}$, $M_- \in \mathbb{R}^{p\times\ell_- p}$, $M_0 \in \mathbb{R}^{p\times p}$ such that

$$\begin{bmatrix} -P_{22} & \begin{bmatrix} 0 & \rho^{-1}Q_A + m\rho^{-1}Q_B C & Q_B \end{bmatrix} \\ \star & U(P, M_\Delta) \end{bmatrix} \prec 0 \tag{3.77a}$$

$$P = \mathrm{blkdiag}(P_{11}, P_{22}) \tag{3.77b}$$

$$(Q_A - P_{22})D^\dagger = 0 \tag{3.77c}$$

$$\begin{bmatrix} M_- & M_0 & M_+ \end{bmatrix} \in \mathbb{M}(\rho, \ell_-, \ell_-, p). \tag{3.77d}$$

Then, with $A = P_{22}^{-1}Q_A$, $B = P_{22}^{-1}Q_B$, the equilibrium $x^\star = Dz^\star$ is globally robustly exponentially stable against $\mathbf{\Delta}(m, L)$ with rate ρ for (3.15). •

Proof. The result follows directly from Theorem 7 noting that (3.77a), (3.73), (3.77d) imply (3.55a), (3.77a) implies $P_{22} \succ 0$ and thus nominal exponential stability by Lemma 9, and (3.77c) together with the choice of C, $D^\dagger$ implies (3.4). □

We note that the latter result provides a convex solution to the original design problem introduced in Section 3.1; however, as it turns out in numerical examples, the restrictions on P are conservative, and we trade off convexity for slower convergence rates.

When it comes to the synthesis problem with additional performance specifications, similar restrictions on P_{p} can be employed. Building upon that, the following Theorem provides a convex solution to this extended problem. A proof is provided in Appendix B.2.10.

Theorem 10. Let $L \geq m > 0$ as well as $n \in \mathbb{N}_{>0}$, $p \in \mathbb{N}_{>0}$ be given. Let $C \in \mathbb{R}^{p\times np}$, $D \in \mathbb{R}^{np\times p}$, $D^\dagger \in \mathbb{R}^{p\times np}$ be defined as in (3.68). Fix $\rho \in (0,1)$. Let some $\ell_-, \ell_+ \in \mathbb{N}$ be given and let M_Δ, $\mathbb{M}$ be defined according to (3.52), (3.42). Let further $B_{\mathrm{p}} = B$, $C_{\mathrm{p}} = C$, $D_{\mathrm{p}} = 0$. Suppose there exist $P_{11} = P_{11}^\top \in \mathbb{R}^{p(\ell_-+\ell_+)\times p(\ell_-+\ell_+)}$, $P_{22} = P_{22}^\top \in \mathbb{R}^{np\times np}$, $Q_A \in \mathbb{R}^{np\times np}$, $Q_B \in \mathbb{R}^{np\times p}$, $P_{\mathrm{p},11} = P_{\mathrm{p},11}^\top \in \mathbb{R}^{p(\ell_-+\ell_+)\times p(\ell_-+\ell_+)}$, $Z = Z^\top \in \mathbb{R}^{p\times p}$, $M_+, M_{\mathrm{p},+} \in \mathbb{R}^{p\times \ell_- p}$, $M_-, M_{\mathrm{p},-} \in \mathbb{R}^{p\times \ell_- p}$, $M_0, M_{\mathrm{p},0} \in \mathbb{R}^{p\times p}$ such that (3.77) holds and

$$\begin{bmatrix} -P_{22} & \begin{bmatrix} 0 & Q_A + mQ_BC & Q_B \end{bmatrix} \\ \star & \mathcal{U}_{\mathrm{p}}(P_{\mathrm{p}}, M_{\mathrm{p},\Delta}) \end{bmatrix} \prec 0 \tag{3.78a}$$

$$P_{\mathrm{p}} = \mathrm{blkdiag}(P_{\mathrm{p},11}, P_{22}) \tag{3.78b}$$

$$\begin{bmatrix} P_{22} & Q_B \\ \star & Z \end{bmatrix} \succ 0 \tag{3.78c}$$

$$\mathrm{tr}(Z) < \gamma^2, \tag{3.78d}$$

$$\begin{bmatrix} M_{\mathrm{p},-} & M_{\mathrm{p},0} & M_{\mathrm{p},+} \end{bmatrix} \in \mathbb{M}(1, \ell_-, \ell_-, p), \tag{3.78e}$$

where $M_{\mathrm{p},\Delta}$ is defined as M_Δ in (3.52) replacing M_+, M_-, M_0 by their counterparts $M_{\mathrm{p},+}$, $M_{\mathrm{p},-}$, $M_{\mathrm{p},0}$. Then, with $A = P_{22}^{-1}Q_A$, $B = P_{22}^{-1}Q_B$, the equilibrium $x^\star = Dz^\star$ is globally robustly exponentially stable against $\mathbf{\Delta}(m, L)$ with rate ρ for (3.15) and (3.15) achieves a robust H_2-performance level of γ against $\mathbf{\Delta}(m, L)$. •

3.3.2.2 Synthesis based on BMI optimization techniques

While Theorem 9, Theorem 10 provide a convex synthesis procedure, the block diagonal structure of P as well as the assumptions on P_{p} are restrictive and result in convergence rates inferior to what can be achieved, see Section 3.3.2.3. As an alternative, BMI optimization techniques, which directly try to solve (3.72), (3.74), (3.76) or variants thereof, can be employed. However, while many of the approaches perform reasonably well as we will also illustrate in the subsequent example section, the non-convex nature of the problem does not allow for any guarantees of finding the global optimizer in general. Here, we employ the so-called alternating method which alternates between solving two different semi-definite programs obtained from fixing two subsets of the set of all decision variables in the BMI. In particular, assuming that $C, D, D^{\dagger}$ are chosen as in (3.68), we alternate between finding the best algorithm A, B in terms of ρ and γ for fixed P, P_{p} and solving for P, P_{p} for this A, B, ρ, γ. Building upon an iterative scheme, having a good initial feasible solution is key for successfully applying this procedure. As it turns out in numerical examples, certain types of parametrized algorithms are better suited for initialization than making use of Theorem 9. In particular, we suppose that the matrices A, B are parametrized by the free parameters $K_i \in \mathbb{R}^{p\times p}, i = 1, 2, \ldots, n$, as

$$A = A_1 + I_{np} + B_1 \begin{bmatrix} 0 & K_2 & \ldots & K_n \end{bmatrix}, \qquad B = B_1 K_1, \tag{3.79}$$

where $A_1 \in \mathbb{R}^{np\times np}$, $B_1 \in \mathbb{R}^{np\times p}$ are defined as

$$A_1 = \begin{bmatrix} 0 & I_p & & & \\ 0 & 0 & I_p & & \\ \vdots & & & \ddots & \\ \vdots & & & & I_p \\ 0 & \ldots & & & 0 \end{bmatrix}, B_1 = \begin{bmatrix} 0 \\ 0 \\ \vdots \\ \vdots \\ I_p \end{bmatrix}. \tag{3.80}$$

The particular structure is motivated by Michalowsky and Ebenbauer (2014), Michalowsky and Ebenbauer (2016) and corresponds to a Euler discretization of the nth order heavy ball method presented in the latter references. For the initialization of the BMI iteration we then fix C, D as in (3.68) and utilize Lemma 12 to find suitable parameters K_i, $i = 1, 2, \ldots, n$, see Section 3.5.2.

3.3.2.3 Numerical results

In this section we employ the presented results to design novel optimization algorithms. We first discuss the case where no performance channel is present and the goal is to only optimize the guaranteed convergence rate. When using Theorem 9, the resulting convergence rates are the same as the ones of the Gradient Descent method, see Table 3.1; also, the algorithms are structurally equivalent to a Gradient Descent. Still, the approach is valuable when considering structured optimization problems, see the numerical results in Section 3.4.2. Employing BMI optimization techniques, the convergence rates of the Triple Momentum Method can be recovered but not improved.

Following the analysis from Section 3.3.1.3, we additionally consider the design of algorithms that are insensitive to additive noise. The numerical results obtained from following the BMI optimization procedure are depicted in Figure 3.8. As already observed in Section 3.3.1.3, there exists a trade-off between convergence rates and robustness towards noise. Further, higher order algorithms can yield better H_2-performances while providing the same convergence rate guarantees. The convex approach from Theorem 10 might yield even better performance bounds; however, the approach is limited to convergence rates worse than the one of the gradient descent algorithm.

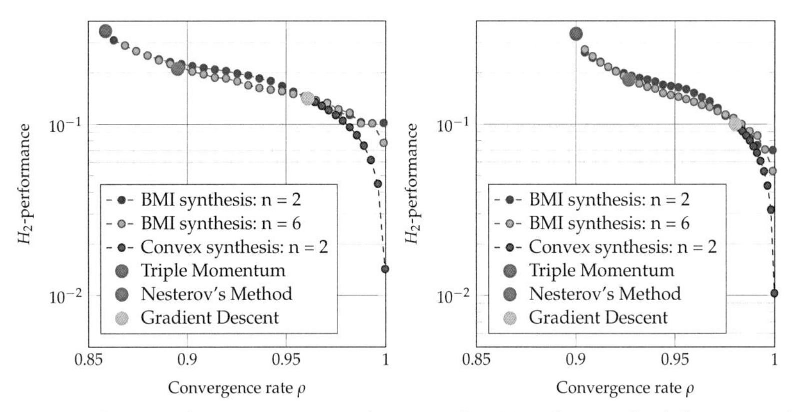

Figure 3.8. Guaranteed convergence rates and corresponding H_2-performance levels for $m = 1$ and $L = 50$ (left) as well as $L = 100$ (right). For comparison, we also plot the H_2-performance levels obtained from Theorem 8 for Nesterov's method as well as the Gradient Descent algorithm.

3.4 Structure Exploiting Algorithms

Up to now the only assumption on the objective function was that $H \in \mathcal{S}_{m,L}$. However, in many situations it is possible to further characterize the objective function in terms of its structure. In such cases, it is to be expected that a tailored algorithm that exploits these properties has much better guaranteed convergence rates than a standard algorithm for the class $\mathcal{S}_{m,L}$. In this section we elaborate on how additional structural knowledge can be incorporated in the presented framework. To this end, for any $L \geq m > 0$, we define the following two sets of functions:

$$\begin{aligned}\mathcal{S}_{m,L}^{\text{n-rep}} = \{H \in \mathcal{S}_{m,L} \mid \exists \phi_i : \mathbb{R} \to \mathbb{R}, \phi_i \in \mathcal{C}^0, i = 1,2,\ldots,p, \text{ such that} \\ \nabla H(x) = \begin{bmatrix}\phi_1(x_1) & \ldots & \phi_p(x_p)\end{bmatrix}^\top \\ \text{for all } x = \begin{bmatrix}x_1 & x_2 & \ldots & x_p\end{bmatrix}^\top \in \mathbb{R}^p\}\end{aligned} \tag{3.81}$$

$$\begin{aligned}\mathcal{S}_{m,L}^{\text{rep}} = \{H \in \mathcal{S}_{m,L} \mid \exists \phi : \mathbb{R} \to \mathbb{R}, \phi \in \mathcal{C}^0, \text{ such that} \\ \nabla H(x) = \begin{bmatrix}\phi(x_1) & \ldots & \phi(x_p)\end{bmatrix}^\top \\ \text{for all } x = \begin{bmatrix}x_1 & x_2 & \ldots & x_p\end{bmatrix}^\top \in \mathbb{R}^p\}.\end{aligned} \tag{3.82}$$

Note that $\mathcal{S}_{m,L}^{\text{n-rep}}$ is the set of all functions with a diagonal Hessian and $\mathcal{S}_{m,L}^{\text{rep}}$ is the set of all functions with a diagonally repeated Hessian. Objective functions of this form appear, e.g., in distributed optimization problems where the objective function is a sum of the agents' individual objective functions. We discuss how this information can be translated to IQCs in Section 3.4.1. Additionally, in Section 3.4.2, we consider parametrized objective functions consisting of a known quadratic and a specifically structured unknown part.

3.4.1 IQCs for the classes $\mathcal{S}_{m,L}^{\text{n-rep}}, \mathcal{S}_{m,L}^{\text{rep}}$

In the following we want to derive IQCs for uncertainties of the form (3.13) and the transformed version (3.25) thereof under the additional assumption that the objective function H is in $\mathcal{S}_{m,L}^{\text{n-rep}}$ or $\mathcal{S}_{m,L}^{\text{rep}}$, respectively. At this point, we benefit from embedding our approach in the standard IQC framework rendering the following derivation straightforward. In the spirit of our previous discussions, we define the following sets of uncertainties

$$\mathbf{\Delta}_{\text{n-rep}}(m,L) = \{\Delta_H \mid H \in \mathcal{S}_{m,L}^{\text{n-rep}}\}, \tag{3.83a}$$

$$\mathbf{\Delta}_{\text{rep}}(m,L) = \{\Delta_H \mid H \in \mathcal{S}_{m,L}^{\text{rep}}\}, \tag{3.83b}$$

with $\Delta_H(y)_t := \nabla H(y_t + z^\star) - my_t$, see (3.13). Observe that $\mathbf{\Delta}_{\text{rep}}(m,L) \subseteq \mathbf{\Delta}_{\text{n-rep}}(m,L) \subseteq \mathbf{\Delta}(m,L)$. Operators of this type have as well been studied in the literature. In fact, the following modification of Lemma 6 holds, see Fetzer and Scherer (2017).

Lemma 10. Let $L \geq m > 0$ and $\ell_-, \ell_+ \in \mathbb{N}$ be given. Let $\mathbf{\Delta}_{\text{n-rep}}(m,L), \mathbf{\Delta}_{\text{rep}}(m,L)$ be defined according to (3.83a), (3.83b) and let $M : \ell_2^p \to \ell_2^p$ be a bounded linear operator defined as

$$M = \operatorname{Toep}(\begin{bmatrix} M_{-\ell_-} & M_{-\ell_-+1} & \cdots & M_{\ell_+} \end{bmatrix}), \tag{3.84}$$

where $M_i \in \mathbb{R}^{p\times p}$. Then (3.32) persists to hold

(a) for all $\Delta \in \mathbf{\Delta}_{\text{n-rep}}(m,L)$ if, for $i \in \{-\ell_-, \ldots, \ell_+\}$,

$$M_i = \operatorname{blkdiag}(m_{i,1}, m_{i,2}, \ldots, m_{i,p}) \tag{3.85}$$

and, for each $k = 1, 2, \ldots, p$, the operator

$$\operatorname{Toep}(\begin{bmatrix} m_{-\ell_-,k} & \cdots & m_{0,k} & \cdots & m_{\ell_+,k} \end{bmatrix}) \tag{3.86}$$

is doubly hyperdominant.

(b) for all $\Delta \in \mathbf{\Delta}_{\text{rep}}(m,L)$ if M is doubly hyperdominant. •

By our discussions from Section 3.2.4, it is not difficult to derive IQCs for exponential stability analysis. We first define the corresponding sets of transformed uncertainties as

$$\mathbf{\Delta}_{\text{n-rep},\rho}(m,L) = \{\rho_- \circ \Delta \circ \rho_+ \mid \Delta \in \mathbf{\Delta}_{\text{n-rep}}(m,L)\} \tag{3.87a}$$

$$\mathbf{\Delta}_{\text{rep},\rho}(m,L) = \{\rho_- \circ \Delta \circ \rho_+ \mid \Delta \in \mathbf{\Delta}_{\text{rep}}(m,L)\}. \tag{3.87b}$$

Following the same steps, again we demand $M_\rho = \rho_- \circ M \circ \rho_-$ to adhere to condition (a) in Lemma 10 if $\Delta_\rho \in \mathbf{\Delta}_{\text{n-rep},\rho}(m,L)$ or to (b) if $\Delta_\rho \in \mathbf{\Delta}_{\text{rep},\rho}(m,L)$. The sets of admissible

$[M_i]_{i\in\{-\ell_-,\dots,\ell_+\}}$ are then defined as

$$\mathbb{M}_{\text{rep}}(\rho,\ell_+,\ell_-,p) = \{ \begin{bmatrix} M_{-\ell_-} & M_{-\ell_-+1} & \dots & M_{\ell_+} \end{bmatrix} \mid M_i \in \mathbb{R}^{p\times p}, M_i \le 0 \text{ for all } i \ne 0, \\ (\textstyle\sum_{i=-\ell_-}^{\ell_+} M_i\rho^{-i})\mathbf{1} \ge 0, \\ \mathbf{1}^\top(\textstyle\sum_{i=-\ell_-}^{\ell_+} M_i\rho^{i}) \ge 0\} \tag{3.88}$$

in the repeated case and as

$$\mathbb{M}_{\text{n-rep}}(\rho,\ell_+,\ell_-,p) = \{ \begin{bmatrix} M_{-\ell_-} & M_{-\ell_-+1} & \dots & M_{\ell_+} \end{bmatrix} \mid M_i = \operatorname{blkdiag}(m_{i,1}, m_{i,2}, \dots, m_{i,p}), \\ M_i \le 0 \text{ for all } i \ne 0, \\ (\textstyle\sum_{i=-\ell_-}^{\ell_+} M_i\rho^{-i})\mathbf{1} \ge 0, \\ \mathbf{1}^\top(\textstyle\sum_{i=-\ell_-}^{\ell_+} M_i\rho^{i}) \ge 0\} \tag{3.89}$$

for the non-repeated case. The corresponding Zames-Falb multipliers $\mathbf{\Pi}^p_{\Delta,\rho,\text{n-rep}}(m,L)$ and $\mathbf{\Pi}^p_{\Delta,\rho,\text{rep}}(m,L)$ are obtained from (3.47) simply by replacing $\mathbb{M}$ by $\mathbb{M}_{\text{n-rep}}(\rho,\ell_+,\ell_-,p)$ and $\mathbb{M}_{\text{rep}}(\rho,\ell_+,\ell_-,p)$, respectively. Analogously to Theorem 6, we then have the following result.

Theorem 11. Let $L \ge m > 0$ and let $\mathbf{\Delta}_{\text{n-rep},\rho}(m,L)$, $\mathbf{\Delta}_{\text{rep},\rho}(m,L)$ be defined as in (3.87a), (3.87b). Then, for each $\rho \in (0,1]$, Δ_ρ satisfies the IQC defined by Π

1. for each $\Delta_\rho \in \mathbf{\Delta}_{\text{n-rep},\rho}(m,L)$ and each $\Pi \in \mathbf{\Pi}^p_{\Delta,\rho,\text{n-rep}}(m,L)$;
2. for each $\Delta_\rho \in \mathbf{\Delta}_{\text{rep},\rho}(m,L)$ and each $\Pi \in \mathbf{\Pi}^p_{\Delta,\rho,\text{rep}}(m,L)$. •

It is then straightforward to include additional structural assumptions in the results from Section 3.3. In a nutshell, we only need to replace the constraint on $\begin{bmatrix} M_- & M_0 & M_+ \end{bmatrix}$ by the respective structured counterpart.

3.4.2 Parametrized objective functions

In the following we assume that the gradient of the objective function $H : \mathbb{R}^p \to \mathbb{R}$ admits the form

$$\nabla H(z) = H_1 z + T^\top \nabla H_2(Tz), \tag{3.90}$$

where $H_1 \in \mathbb{R}^{p\times p}$, $H_1 \succ 0$, $T \in \mathbb{R}^{q\times p}$ are known whereas $H_2 : \mathbb{R}^q \to \mathbb{R}$ is unknown but fulfills $H_2 \in \mathcal{S}_{m_2,L_2}$ for some known constants $L_2 \ge m_2 > 0$. Objective functions of this specific form arise in the context of linear relaxed logarithmic barrier function based model predictive control (Feller & Ebenbauer, 2017a), where fast converging optimization algorithms are crucial for the practical applicability of the control scheme. It is clear that $H \in \mathcal{S}_{m,L}$ with $m = \lambda_{\min}(H_1)$, $L = \lambda_{\max}(H_1 + T^\top T L_2)$, where $\lambda_{\min}$, $\lambda_{\max}$ denote the minimal and maximal eigenvalue, respectively. Again, we denote by $z^\star$ the minimizer of

H. Note that m, L can both be computed under the assumption that H_1 and T are known. Hence, we can make use of any algorithm for the class $\mathcal{S}_{m,L}$; in particular, we can make use of the fastest known algorithm in the class of considered algorithms, i.e., the Triple Momentum Method (Van Scoy et al., 2018). However, with the structure of H being known and taking the form (3.90), it is possible to obtain algorithms with improved convergence rate guarantees in many cases using the presented framework.

To this end, consider an algorithm of the form (3.2) under the structural assumption (3.90) which then takes the form

$$x_{t+1} = (A + BH_1C)x_t + BT^\top \nabla H_2(TCx_t) \tag{3.91a}$$
$$z_t = Dx_t. \tag{3.91b}$$

By the same state transformation $\xi_t = x_t - D^\dagger z^\star$ and under the assumptions (3.4) we then obtain the transformed dynamics

$$\xi_{t+1} = (A + BH_1C)\xi_t + BT^\top \nabla H_2(TC\xi_t + Tz^\star) + BH_1 z^\star \tag{3.92a}$$
$$z_t = D\xi_t + z^\star. \tag{3.92b}$$

Since $z^\star$ is the minimizer of H, we have $\nabla H(z^\star) = H_1 z^\star + T^\top \nabla H_2(Tz^\star) = 0$. Further, with $H_1 \succeq mI \succ 0$, H_1 is non-singular and hence the previous equality implies that $z^\star = -H_1^{-1}T^\top \nabla H_2(Tz^\star)$. Using this in (3.92), we obtain

$$\xi_{t+1} = (A + BH_1C)\xi_t + BT^\top \left(\nabla H_2(TC\xi_t + Tz^\star) - \nabla H_2(Tz^\star)\right) \tag{3.93a}$$
$$z_t = D\xi_t + z^\star. \tag{3.93b}$$

In the same spirit as in the unstructured case, we hence write the feedback interconnection in the standard form as

$$\xi_{t+1} = (A + BH_1C + m_2 BT^\top TC)\xi_t + w_t \tag{3.94a}$$
$$y_t = TC\xi_t \tag{3.94b}$$
$$w_t = \Delta_{H_2}(y)_t \tag{3.94c}$$

with the new uncertainty $\Delta_{H_2} : \ell_2^q \to \ell_2^q$, defined as

$$\Delta_{H_2}(y)_t = \nabla H_2(y_t + Tz^\star) - \nabla H_2(Tz^\star) - m_2 y_t \tag{3.95}$$

for any $y = [\ldots, y_0, y_1, y_2, \ldots] \in \ell_2^q$ and the corresponding class of uncertainties

$$\mathbf{\Delta} = \{\Delta_{H_2} : H_2 \in \mathcal{S}_{0,L_2-m_2}\}. \tag{3.96}$$

Note that, again, the so-defined $\Delta \in \mathbf{\Delta}$ is a slope-restricted operator in the sector $[0, L_2 - m_2]$. Hence we can make use of all the previously derived IQCs for the new uncertainty. In a nutshell, this means that we can employ the very same techniques with the substitutions $A + mBC \mapsto A + BH_1C + m_2BT^\top TC$, $B \mapsto BT^\top$, $C \mapsto TC$, $L - m \mapsto L_2 - m_2$. We emphasize that this also applies to the convex synthesis procedures presented in Section 3.3.2.1 using corresponding substitutions for Q_A, Q_B.

In the following numerical example we show that exploiting structural knowledge about the objective function can lead to a significant improvement in convergence rate guarantees. We choose $H_1 = \text{blkdiag}(1, 2, 10, 4)$, i.e., $m = 1$, and further let T be given by

$$T = \begin{bmatrix} 2 & -7 & 0 & 5 \\ -1 & 4 & -3 & 2 \\ 0 & -2 & 1 & 0 \end{bmatrix} \tag{3.97}$$

$m_2 = 1$, and let L_2 vary from 1 to 20, resulting in L ranging from 89.677 to 1774.5. For the problem at hand, we design tailored algorithms based on the convex synthesis procedure from Section 3.3.2.1 and the BMI optimization approach described in Section 3.3.2.2. The convergence rates of the respective algorithms are depicted in Figure 3.9. In the considered example, structure exploiting algorithms provide much better guaranteed convergence rates compared to the fastest known standard algorithm, the Triple Momentum Method, using only m and L as known parameters but neglecting any structural properties of H. Still, we emphasize that the achievable improvements heavily depend on the specific problem. The numerical results also show that additionally assuming that $H_2 \in \mathcal{S}_{m,L}^{\text{rep}}$ and employing the results from Section 3.4.1 can further improve the guaranteed convergence rates.

3.5 Discussions

In this section we discuss possible extensions of the presented framework as well as related approaches. We do not aim for working out the extensions in detail but rather give suggestions how to approach them.

3.5.1 Constrained Optimization

In the following we discuss how the presented framework can be extended to constrained optimization problems. To this end, we propose two possible approaches: (i) using (relaxed) barrier functions to reformulate the constrained optimization problem as an unconstrained one, and (ii) employing saddle-point dynamics as introduced in Section 2.2.

Barrier Function Based Approach. Barrier functions have successfully been utilized in convex optimization for a long time, in particular in interior-point methods (Nesterov & Nemirovskii, 1994). The idea of barrier function based methods is to incorporate inequality constraints in optimization problems into the objective function by means of a suitably defined barrier function, i.e., a function that, ideally, is zero inside the feasible set defined by the inequality constraints and infinity outside. Several smooth approximations have been proposed, where logarithmic barrier functions are most widely used which we will also concentrate on in the following. To this end, consider a constrained convex optimization problem (1.1) without equality constraints (i.e., $a(z) = 0$) and suppose that $H \in \mathcal{S}_{m,L}$

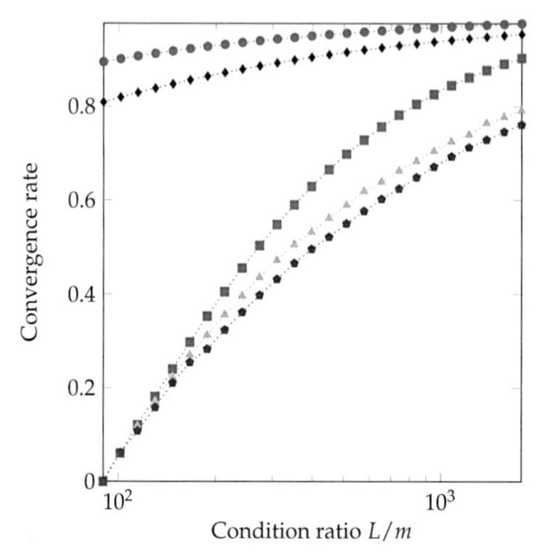

Figure 3.9. Convergence rate guarantees provided by structure exploiting algorithms designed for the example from Section 3.4.2 using the conservative convex approach (■), the BMI optimization approach (▲), as well as the BMI optimization approach under the additional assumption that $H_2 \in \mathcal{S}_{m,L}^{\text{rep}}$ (⬟). For comparison, we plot the convergence rates provided by the Triple Momentum Method (●) and the lower bound for any first order algorithm when the structure is neglected (♦).

and Assumption 1 holds. The logarithmic barrier function associated to this problem is then given by

$$B(z) = -\sum_{i=1}^{n_{\text{ineq}}} \ln\big(-c_i(z)\big), \tag{3.98}$$

where $\ln : \mathbb{R}_{>0} \to \mathbb{R}$ is the natural logarithm. Barrier function based methods then try to solve the unconstrained optimization problem

$$\underset{z \in \mathbb{R}^p}{\text{minimize}} \quad \tilde{H}(z) = H(z) + \varepsilon B(z) \tag{3.99}$$

where $\varepsilon > 0$ is a sufficiently small parameter. Intuitively, if ε tends to zero, then the modified objective function $\tilde{H}$ converges pointwise to the original objective function H for all $z \in \mathbb{R}^p$ with $c(z) < 0$. However, such barrier functions are undefined if $c_i(z) \geq 0$ for some $i \in \{1, 2, \dots, n_{\text{ineq}}\}$, which is why so-called *relaxed logarithmic barrier functions* have been introduced (Ben Tal, Tsibulevskii, & Yusefovich, 1992; Hauser & Saccon, 2006; Nash, Polyak, & Sofer, 1994) that smoothly extend classical barrier functions by a suitable penalty term for all $z \in \mathbb{R}^p$ where $c(z) \geq 0$. In the following we employ a special class of relaxed logarithmic barrier functions with the property that the barrier function is itself a function in the class

$\mathcal{S}_{\tilde{m},\tilde{L}}$ for some $\tilde{L} > \tilde{m} > 0$, given that the inequality constraints are affine. More precisely, we consider quadratic relaxations

$$\beta_\delta(z) = \tfrac{1}{2}\big(\big(\tfrac{z-2\delta}{\delta}\big)^2 - 1\big) - \ln(\delta), \tag{3.100}$$

with $\delta > 0$. This yields the globally defined relaxed barrier function $\hat{B} : \mathbb{R}^p \to \mathbb{R}$ given by

$$\hat{B}(z) = \sum_{i=1}^{n_{\text{ineq}}} \hat{B}_i(z) \tag{3.101a}$$

$$\hat{B}_i(z) = \begin{cases} -\ln\big(-c_i(z)\big) & \text{if } c_i(z) < -\delta \\ \beta_\delta\big(-c_i(z)\big) & \text{if } c_i(z) \geq -\delta. \end{cases} \tag{3.101b}$$

as well as the corresponding optimization problem

$$\underset{z\in\mathbb{R}^p}{\text{minimize}} \quad \hat{H}(z) = H(z) + \varepsilon\hat{B}(z). \tag{3.102}$$

Relaxed barrier functions of this type have been proposed and investigated in the context of model predictive control (Feller & Ebenbauer, 2017a), where also the role of the barrier function parameters ε, δ has been analyzed. We do not go into detail here, but consider ourselves satisfied with the statement that the modified unconstrained problem (3.102) will yield an approximate solution to the original constrained optimization problem for sufficiently small ε, δ. More interestingly for the present application, it has been shown in Feller and Ebenbauer (2017b) that for affine inequality constraints

$$c(z) = C_{\text{ineq}}z - d_{\text{ineq}}, \tag{3.103}$$

where $C_{\text{ineq}} \in \mathbb{R}^{n_{\text{ineq}}\times p}$, $d_{\text{ineq}} \in \mathbb{R}^{n_{\text{ineq}}}$. the modified objective function $\hat{H}$ is in the class $\mathcal{S}_{\tilde{m},\tilde{L}}$ with parameters $\tilde{m} = m$ and

$$\tilde{L} = \max_{z\in\mathbb{R}^p} \big\{\lambda_{\max}\big(\nabla^2 H(z) + \tfrac{\varepsilon}{\delta^2} z^\top C_{\text{ineq}}^\top C_{\text{ineq}} z\big)\big\}, \tag{3.104}$$

where $\lambda_{\max}(P)$ denotes the maximal eigenvalue of a symmetric matrix P. In particular, since $H \in \mathcal{S}_{m,L}$, an upper bound for $\tilde{L}$ is given by

$$\tilde{L} \leq L + \lambda_{\max}\big(\tfrac{\varepsilon}{\delta^2} C_{\text{ineq}}^\top C_{\text{ineq}}\big). \tag{3.105}$$

We may hence utilize the methods developed in the present chapter to design algorithms to solve (3.102). In particular, for quadratic objective functions $H(z) = z^\top H_1 z$, $H_1 \in \mathbb{R}^{p\times p}$, $H_1 \succ 0$, the gradient of the modified objective function $\nabla\hat{H}$ is parametrized as in (3.90) and we can employ the results from Section 3.4.2.

Analysis and Design of Saddle-Point Dynamics. In this paragraph, we consider constrained optimization problems of the form (1.1) without inequality constraints (i.e., $c(z) = 0$), where $H \in \mathcal{S}_{m,L}$ and Assumption 1 holds. We note that affine inequality constraints can be incorporated following a barrier function based reformulation as discussed in the latter paragraph. Motivated by continuous-time saddle-point dynamics as presented in Section 2.2, we consider algorithms of the form

$$\begin{aligned} \mathbf{z}_{t+1} &= A_z \mathbf{z}_t + B_z \nabla_z L(C_z x_t, C_\nu \boldsymbol{\nu}_t) && (3.106a)\\ \boldsymbol{\nu}_{t+1} &= A_\nu \boldsymbol{\nu}_t + B_\nu \nabla_\nu L(C_z x_t, C_\nu \boldsymbol{\nu}_t) && (3.106b)\\ z_t &= D_z \mathbf{z}_t && (3.106c)\\ \nu_t &= D_\nu \boldsymbol{\nu}_t, && (3.106d) \end{aligned}$$

where $L(z,\nu) = H(z) + \nu^\top a(z)$ is the Lagrangian associated to (1.1). Given $L > m > 0$, the affine equality constraint

$$a(z) = A_{\text{eq}} z - b_{\text{eq}} \tag{3.107}$$

and $n, n_\nu \in \mathbb{N}_{>0}$, our goal is then to design $A_z \in \mathbb{R}^{np \times np}$, $B_z \in \mathbb{R}^{np \times p}$, $C_z \in \mathbb{R}^{p \times np}$, $A_\nu \in \mathbb{R}^{n_\nu n_{\text{eq}} \times n_\nu n_{\text{eq}}}$, $B_\nu \in \mathbb{R}^{n_\nu n_{\text{eq}} \times n_{\text{eq}}}$, $C_\nu \in \mathbb{R}^{n_{\text{eq}} \times n_\nu n_{\text{eq}}}$, $D_z \in \mathbb{R}^{p \times np}$, $D_\nu \in \mathbb{R}^{n_{\text{eq}} \times n_\nu n_{\text{eq}}}$ such that (z_t, ν_t) converges to a saddle point (cf. Definition 2) of the Lagrangian L for all $H \in \mathcal{S}_{m,L}$, thereby providing a solution to the constrained optimization problem (1.1). We note that (3.106) can be interpreted as the discrete-time counterpart of the higher order saddle-point dynamics (2.122). As discussed in Section 2.2, the existence of a saddle point of L is ensured by Assumption 1; for uniqueness we additionally assume that the constraint qualification Assumption 2 holds, i.e., A_{eq} has full rank. We then denote by $(z^\star, \nu^\star)$ the unique saddle point of L. To address the problem, we follow the same route as in Section 3.2 reformulating the problem in the IQC framework.

To this end, suppose that there exist $D_z^\dagger \in \mathbb{R}^{np \times p}$, $D_\nu^\dagger \in \mathbb{R}^{n_\nu n_{\text{eq}} \times n_{\text{eq}}}$ such that

$$\begin{aligned} D_z D_z^\dagger &= I_p & D_\nu D_\nu^\dagger &= I_{n_{\text{eq}}} && (3.108a)\\ C_z D_z^\dagger &= I_p & C_\nu D_\nu^\dagger &= I_{n_{\text{eq}}} && (3.108b)\\ (A_z - I) D_z^\dagger &= 0_{np \times p} & (A_\nu - I) D_\nu^\dagger &= 0_{n_\nu n_{\text{eq}} \times n_{\text{eq}}}. && (3.108c) \end{aligned}$$

These conditions are motivated the necessary conditions derived in Section 3.2.1; we do not prove necessity here but it is expected that a result similar to Theorem 3 can be derived. Next, consider the state transformation $\xi_t = \mathbf{z}_t - D_z^\dagger z^\star$, $\mu_t = \boldsymbol{\nu}_t - D_\nu^\dagger \nu^\star$. Utilizing (3.108) as well as the Karush-Kuhn-Tucker conditions fulfilled by the saddle point $(z^\star, \nu^\star)$

$$a(z^\star) = A_{\text{eq}} z^\star - b_{\text{eq}} = 0, \qquad \nabla_x L(z^\star, \nu^\star) = \nabla H(z^\star) + A_{\text{eq}}^\top \nu^\star = 0, \tag{3.109}$$

the transformed dynamics can be expressed as

$$\begin{bmatrix} \xi_{t+1} \\ \mu_{t+1} \end{bmatrix} = \begin{bmatrix} A_z & B_z A_{\text{eq}}^\top C_\nu \\ B_\nu A_{\text{eq}} C_z & A_\nu \end{bmatrix} \begin{bmatrix} \xi_t \\ \mu_t \end{bmatrix} + \begin{bmatrix} B_z \\ 0 \end{bmatrix} \left(\nabla H(C_z \xi_t + z^\star) - \nabla H(z^\star) \right). \tag{3.110}$$

Again, we interpret ∇H as the uncertainty and define the bounded and causal uncertain operator $\Delta_H : \ell_e^p \to \ell_e^p$ as

$$\Delta_H(y)_t := \nabla H(y_t + z^\star) - \nabla H(z^\star) - m y_t \tag{3.111}$$

for any $y = [y_0, y_1, y_2, \dots] \in \ell_e^p$. We note that this is the same as (3.13) but therein, $\nabla H(z^\star) = 0$. We then let the corresponding set of uncertainties $\mathbf{\Delta}(m, L)$ be defined as in (3.14). In the spirit of Section 3.2.2, we then consider the following feedback interconnection

$$\begin{aligned} x_{t+1} &= (A_{\mathrm{SP}} + m B_{\mathrm{SP}} C_{\mathrm{SP}}) x_t + B_{\mathrm{SP}} w_t && (3.112a)\\ y_t &= C_{\mathrm{SP}} x_t && (3.112b)\\ w_t &= \Delta_H(y_t), && (3.112c) \end{aligned}$$

where $x_t = \begin{bmatrix} \xi_t^\top & \mu_t^\top \end{bmatrix}^\top$ and

$$A_{\mathrm{SP}} = \begin{bmatrix} A_z & B_z A_{\mathrm{eq}}^\top C_v \\ B_v A_{\mathrm{eq}} C_z & A_v \end{bmatrix}, \quad B_{\mathrm{SP}} = \begin{bmatrix} B_z \\ 0 \end{bmatrix}, \quad C_{\mathrm{SP}} = \begin{bmatrix} C_z & 0 \end{bmatrix}. \tag{3.113}$$

Neglecting the performance channel, the dynamics (3.112) have the very same form as (3.15), we may hence apply the results developed in the present chapter to analyze and design the algorithm (3.112). We emphasize that it is straightforward to extend (3.112) by a performance channel. However, it is important to note that, in contrast to (3.15), here, the matrices $A_{\mathrm{SP}}, B_{\mathrm{SP}}, C_{\mathrm{SP}}$ defining the algorithm are structured according to (3.113). While this does not affect the results from Section 3.3.1 concerning algorithm analysis, some results on algorithm synthesis that are based on variable transformations might not apply.

The algorithm (3.112) explicitly utilizes the equality constraint. Sometimes it is desired to design an algorithm that is applicable to a class of constrained optimization problems (1.1) where not only $H \in \mathcal{S}_{m,L}$ is unknown but also the equality constraint determined by a is only known to be affine. This problem can be treated similarly by redefining the uncertainty. More precisely, reconsider (3.110) and define the uncertain operator $\Delta : \ell_e^p \to \ell_e^{p+n_{\mathrm{eq}}}$ as

$$\Delta\left(\begin{bmatrix} y_z \\ y_v \end{bmatrix}\right)_t := \begin{bmatrix} \Delta_H(y_z)_t + \left(\frac{\partial a}{\partial z}(y_{z,t})\right)^\top y_{v,t} \\ a(y_{z,t} - z^\star) \end{bmatrix} \tag{3.114}$$

for any $y_z = [y_{z,0}, y_{z,1}, y_{z,2}, \dots] \in \ell_e^p$, $y_v = [y_{v,0}, y_{v,1}, y_{v,2}, \dots] \in \ell_e^{n_{\mathrm{eq}}}$. Again, the feedback interconnection is then formulated in the standard form as

$$\begin{aligned} x_{t+1} &= (\tilde{A}_{\mathrm{SP}} + \Gamma \tilde{B}_{\mathrm{SP}} \tilde{C}_{\mathrm{SP}}) x_t + \tilde{B}_{\mathrm{SP}} \tilde{w}_t && (3.115a)\\ \tilde{y}_t &= \tilde{C}_{\mathrm{SP}} x_t && (3.115b)\\ \tilde{w}_t &= \Delta(\tilde{y}_t), && (3.115c) \end{aligned}$$

where $x_t = \begin{bmatrix} \xi_t^\top & \mu_t^\top \end{bmatrix}^\top$ and

$$\tilde{A}_{\mathrm{SP}} = \begin{bmatrix} A_z & 0 \\ 0 & A_v \end{bmatrix}, \quad \tilde{B}_{\mathrm{SP}} = \begin{bmatrix} B_z & 0 \\ 0 & B_v \end{bmatrix}, \quad C_{\mathrm{SP}} = \begin{bmatrix} C_z & 0 \\ 0 & C_v \end{bmatrix}, \quad \Gamma = \begin{bmatrix} mI & 0 \\ 0 & 0 \end{bmatrix}. \tag{3.116}$$

To apply the developed framework in this situation as well, we require a suitable IQC for the uncertainty from (3.114). We do not elaborate on that here but note that similar argument as in the previous part should apply since a is affine.

3.5.2 Lyapunov-Based Approaches

In this section we discuss an alternative approach to the problem introduced in Section 3.1 that is also based on a reformulation in the robust control framework but utilizes Lyapunov theory and the S-procedure (see Lemma 18), two tools that might be more familiar than IQCs to some readers. We first briefly explain the core idea of this approach and then apply it to the specific problem of designing parametrized algorithms as introduced and utilized as an initial guess for alternating BMI optimization in Section 3.3.2.2. We only consider convergence rates as an algorithm quality criterion; we emphasize that the approach can be extended accordingly to include quadratic performance measures employing standard results. We note that the basic idea of the following approach has already been employed in Michalowsky and Ebenbauer (2014) in a continuous-time setup.

To explain the idea, consider the transformed algorithm (3.12) neglecting the performance channel. By Lyapunov's direct method, if we manage to find a Lyapunov function V, positive definite with respect to the origin, such that $V(\xi_{t+1}) - V(\xi_k) < 0$ for all non-zero $\xi_k \in \mathbb{R}^{np}$ and all $H \in \mathcal{S}_{m,L}$, then the origin is (robustly) stable for (3.12). Robust exponential stability with rate $\rho \in (0,1)$ can be concluded if

$$V(\xi_{t+1}) - \rho^2 V(\xi_k) < 0 \tag{3.117}$$

for all non-zero $\xi_k \in \mathbb{R}^{np}$ and all $H \in \mathcal{S}_{m,L}$ under the additional requirement

$$\alpha_1 \|\xi\|^2 \leq V(\xi) \leq \alpha_2 \|\xi\|^2 \tag{3.118}$$

for all $\xi \in \mathbb{R}^{np}$ for some $\alpha_1, \alpha_2 \in \mathbb{R}_{>0}$. To see this, note that (3.117) implies that $V(\xi_k) < \rho^{2k} V(\xi_0)$ for all non-zero $\xi_0 \in \mathbb{R}^{np}$ and all $H \in \mathcal{S}_{m,L}$, hence, by (3.118), we infer that $\|\xi_k\|^2 \leq \frac{\alpha_2}{\alpha_1} \rho^{2k} \|\xi_0\|^2$ and robust exponential stability follows by Definition 9. In the following we consider quadratic Lyapunov functions

$$V(\xi) = \xi^\top P \xi \tag{3.119}$$

with $P = P^\top \succ 0$ that trivially fulfill (3.118) with $\alpha_1 = \lambda_{\min}(P)$, $\alpha_2 = \lambda_{\max}(P)$, where $\lambda_{\min}(P), \lambda_{\max}(P)$ denote the minimal and maximal eigenvalue of P, respectively. As it has been observed in Hu and Lessard (2017) for Nesterov's Method in particular, more general Lyapunov functions might yield less conservative results in terms of the convergence rates. Within this respect, we obtained first promising results utilizing Lur'e type Lyapunov functions

$$V_{\text{Lur'e}}(\xi) = \xi^\top P \xi + H(M\xi + z^\star) - H(z^\star), \tag{3.120}$$

where $P \in \mathbb{R}^{np \times np}$, $M \in \mathbb{R}^{p \times np}$ fulfill $\xi^\top P \xi > 0$ for all $\xi \neq 0$ such that $M\xi = 0$, hence rendering $V_{\text{Lur'e}}$ positive definite with respect to the origin. We do not discuss that further but concentrate on quadratic Lyapunov functions (3.119) in the remainder. By (3.12), for quadratic Lyapunov functions, the inequality (3.117) is a quadratic inequality in (ξ_k, w_k) for a suitably defined w_k depending on ∇H. Hence, if we manage to derive another quadratic inequality in (ξ_k, w_k) that is ensured to hold for all $H \in \mathcal{S}_{m,L}$, then we can employ the S-procedure in order to obtain an inequality from (3.117) that is independent of w_k and hence also independent of the unknown term ∇H. We provide such an inequality in the following Lemma; a proof is provided in Appendix B.2.11.

Lemma 11. Let $L > m \geq 0$ be given and suppose that $H \in \mathcal{S}_{m,L}(\mathbb{R}^p)$. For any $z, \bar{z} \in \mathbb{R}^p$, define

$$v(z) = \nabla H(z + \bar{z}) - \nabla H(\bar{z}) - \beta z, \tag{3.121}$$

where $0 \leq \beta \leq \frac{L+m}{2}$. Then

$$\begin{bmatrix} z \\ v(z) \end{bmatrix}^\top \begin{bmatrix} (L-\beta)I_p & 0 \\ 0 & -\frac{1}{L-\beta}I_p \end{bmatrix} \begin{bmatrix} z \\ v(z) \end{bmatrix} \geq 0 \tag{3.122}$$

for all $z, \bar{z} \in \mathbb{R}^p$. •

In the remainder of this section we apply these ideas to design parametrized algorithms. More precisely, we consider algorithms of the form (3.2) where A, B are parametrized by $K_i \in \mathbb{R}^{p \times p}$, $i = 1, 2, \ldots, n$, as in (3.79) and C, D are given by (3.68). We follow the procedure from Michalowsky and Ebenbauer (2014), Michalowsky and Ebenbauer (2016) in a discrete-time setting to embed the design problem into a robust state-feedback synthesis problem and extend it to structured objective functions (3.90). To this end, consider an algorithm of the form (3.2) with A, B as in (3.79) together with the following state transformation

$$\xi_{k,1} := \nabla H(Cx_k), \quad \xi_{k,i} = x_{k,i} \text{ for } i = 2, 3, \ldots, n. \tag{3.123}$$

This yields the transformed dynamics

$$\xi_{t+1} = A_2 \xi_k + B_1 K \xi_k + B_2 \Big(\nabla H(C(A_1 + I)x_k + CB_1 K \xi_k) - \nabla H(Cx_k) \Big), \tag{3.124}$$

where

$$A_2 = \begin{bmatrix} I & 0 & & & \\ 0 & I & I & & \\ \vdots & & I & \ddots & \\ \vdots & & & \ddots & I \\ 0 & \ldots & & & I \end{bmatrix}, B_2 = \begin{bmatrix} I \\ 0 \\ \vdots \\ \vdots \\ 0 \end{bmatrix}. \tag{3.125}$$

The main purpose of this transformation is to embed the problem into a standard robust state-feedback synthesis problem. In view of Section 3.4.2, we additionally suppose that the gradient of the objective function is parametrized as

$$\nabla H(z) = H_1 z + T^\top \nabla H_2(Tz), \tag{3.126}$$

where $H_1 \in \mathbb{R}^{p\times p}$, $T \in \mathbb{R}^{q\times p}$ are known whereas $H_2 : \mathbb{R}^q \to \mathbb{R}$ is unknown but fulfills $H_2 \in \mathcal{S}_{m_2,L_2}$ for some known constants $L_2 \geq m_2 > 0$. Note that this trivially also includes the case when no structural assumptions are taken on the objective function, simply by letting $H_1 = 0$, $T = I$. The following result then follows by using quadratic Lyapunov functions and Lemma 11 for H_2 together with the S-procedure as explained beforehand; more details are given in the proof provided in Appendix B.2.12.

Lemma 12. Let $L_2 \geq m_2 > 0$, $n \in \mathbb{N}_{>0}$, $p \in \mathbb{N}_{>0}$ and $H_1 \in \mathbb{R}^{p\times p}$, $m_1 I \preceq H_1 \preceq L_1 I$, $L_1 \geq m_1 \geq 0$, $T \in \mathbb{R}^{q\times p}, q \in \mathbb{N}_{>0}$, be given. Let $C \in \mathbb{R}^{p\times np}$ be defined as in (3.68). Fix $\rho \in (0,1)$. Suppose there exist $M \in \mathbb{R}^{p\times np}$, $Q \in \mathbb{R}^{np\times np}$, such that

$$\begin{bmatrix} Q & 0 & \bar{A}Q + \bar{B}M & \frac{1}{2}(L_2 - m_2)\bar{G} \\ \star & I & TCA_1 + TCM & 0 \\ \star & \star & \rho^2 Q & 0 \\ \star & \star & \star & I \end{bmatrix} \succ 0, \tag{3.127}$$

where

$$\bar{A} = A_2 + B_2 H_1 C A_1 + \tfrac{1}{2}(L_2 + m_2) B_2 T^\top T C B_1 \tag{3.128a}$$
$$\bar{B} = B_1 + B_2 H_1 C B_1 + \tfrac{1}{2}(L_2 + m_2) B_2 T^\top T C B_1 \tag{3.128b}$$
$$\bar{G} = B_2 T^\top. \tag{3.128c}$$

Then, with $K = MQ^{-1}$ and A, B given by (3.79) and C, D as in (3.68), the equilibrium $x^\star = Dz^\star$ is globally exponentially stable with rate ρ for (3.2) for all H in the form (3.126) with $H_2 \in \mathcal{S}_{m_2,L_2}$. •

The latter result allows us to design parametrized optimization algorithms, both for structured and unstructured objective functions. We mainly make use of this result in the synthesis procedure as explained in Section 3.3.2.2; we note that this result can also be used for discrete-time Extremum Control of linear systems similar to what has been proposed in a continuous-time setting (Michalowsky & Ebenbauer, 2014, 2016).

3.5.3 Continuous-Time Algorithms

While discrete-time algorithms constitute the majority of optimization algorithms, there has been renewed interest in continuous-time algorithms in the last decades (Brockett, 1991; Gharesifard & Cortés, 2014; Helmke & Moore, 1994). Besides providing insight into discrete-time algorithms (Dürr & Ebenbauer, 2012; Su et al., 2014; Wibisono et al., 2016), they also have direct applications to control tasks where the desired equilibrium is characterized by an optimization problem as it is, e.g., in Extremum Control (Michalowsky & Ebenbauer, 2014, 2016; Michalowsky et al., 2017a) or in control approaches based on artificial potential functions such as in formation control or obstacle avoidance (Khatib, 1986; Koren & Borenstein, 1991). As another advantage of embedding the algorithm design problem in the robust control framework which has been established for discrete- as well as continuous-time dynamics, it is expected that the results can be transferred to the continuous-time setting. Exponential convergence results for uncertain continuous-time feedback interconnections using IQCs have already been established in Hu and Seiler (2016). In view of Chapter 2, an extension of the presented methodology to continuous-time dynamics would allow to design tailored distributed optimization algorithms by combining the ideas from Section 3.5.1 with the methodology established in Chapter 2. Within that respect, it is an interesting problem to come up with performance criteria for the non-distributed algorithm that result in a well-performing distributed approximation. As already observed in numerical results in Section 2.5.2 as well as in a related setup in Michalowsky and Ebenbauer (2014), the dynamics that is approximated heavily affects the trajectories of the corresponding approximating dynamics; hence a systematic design procedure could be of great value.

3.6 Summary and Conclusion

We presented a novel and general framework for analyzing and designing robust and structure exploiting optimization algorithms suitable for solving a class of unconstrained optimization problems with strongly convex objective function and possible additional structure. Building upon the well-studied field of robust control theory based on integral quadratic constraints and adapting it to our needs, we provide an approach to the design of robust and structure exploiting optimization algorithms specifically tailored to the class of optimization problems at hand. Several numerical examples illustrate that tailored algorithms designed following the presented methodology can outperform standard algorithms in terms of robustness against noise and guaranteed convergence rates in the considered scenarios.

One key advantage of the approach is that it allows for systematic extensions by suitable adaptations of existing results from robust control theory. In particular, it is to be expected that further characterizations of the class of objective functions as well as performance characterizations, both stated in terms of suitable integral quadratic constraints, can be embedded in the presented framework. We further sketched how the presented framework can be extended to constrained optimization problems utilizing either a barrier function based approach, saddle-point dynamics or a combination thereof. Further research should be carried out into that direction, especially concerning a distributed implementation utilizing the methodology developed in Chapter 2. Therein, as already mentioned beforehand, one major task is to develop meaningful performance specifications that reflect the approximation quality. Another promising application scenario is real-time on-line optimization, in particular barrier function based model predictive control. Optimization problems encountered within this setup are highly structured; not only do they fit into the structure considered in Section 3.4.2 but even more is known, e.g., since parts of the constraints represent the dynamics of the system to be controlled. It is an interesting question in how far this additional knowledge can yield faster algorithms.

4 Summary and Conclusion

In this thesis, we considered the analysis and design of gradient-based optimization algorithms from a systems theoretic perspective. Throughout the thesis, the main premise was to recast the specific problem in a systems and control theoretic framework and employ appropriately extended and modified tools from control theory to systematically address these problems.

In Chapter 2 we designed continuous-time distributed optimization algorithms for a large class of constrained convex optimization problems under mild assumptions on the communication topology. Starting from a non-distributed algorithm, we proposed a two-step approach that allows to systematically derive a distributed approximation thereof. While we focused on saddle-point dynamics as an illustrative example for a – in general situations – non-distributed algorithm capable of solving constrained convex optimization problems, we showed that the proposed methodology applies to a large class of non-distributed algorithms. The core observation that enabled the approach is that parts in the non-distributed algorithm that do not yield a distributed implementation – we called them non-admissible vector fields – can be represented by Lie brackets of admissible vector fields. Utilizing Lie bracket approximation techniques known from geometric control theory, a distributed approximation can then be derived systematically, see Figure 4.1 for a schematic overview of the approach. With respect to the first step of determining admissible Lie bracket representations of non-admissible vector fields, we characterized the class of non-admissible vector fields the approach applies to in terms of the communication graph and discussed its limitations. For the second step of calculating suitable approximating input sequences, based on an algorithm known from literature, we showed how the specific structure of the admissible Lie bracket representations can be exploited to derive a simpler algorithm tailored to the problem at hand.

While we concentrated on distributed implementations of a given algorithm in the first chapter, in Chapter 3 we considered the design of discrete-time gradient-based algorithms for unconstrained convex optimization. We first proposed a class of gradient-based optimization algorithms taking the form of a to be designed linear system in feedback with the gradient of the objective function and formulated necessary conditions for the system to solve the optimization problem, thereby characterizing the class of algorithms. The main observation that links the algorithm design problem to control theory is that the considered algorithms take the form of a Lur'e system, a class of systems well-studied in control

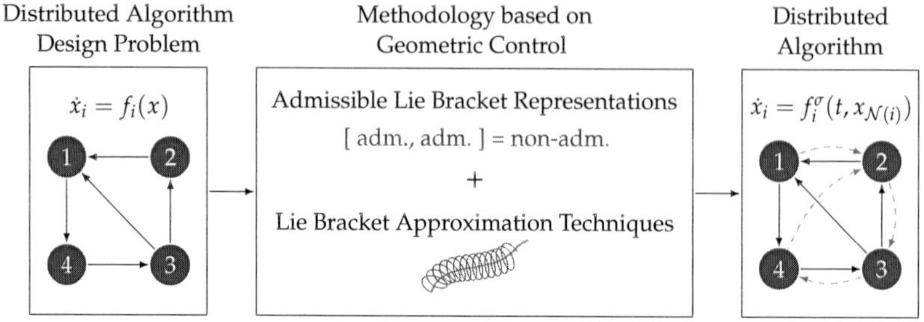

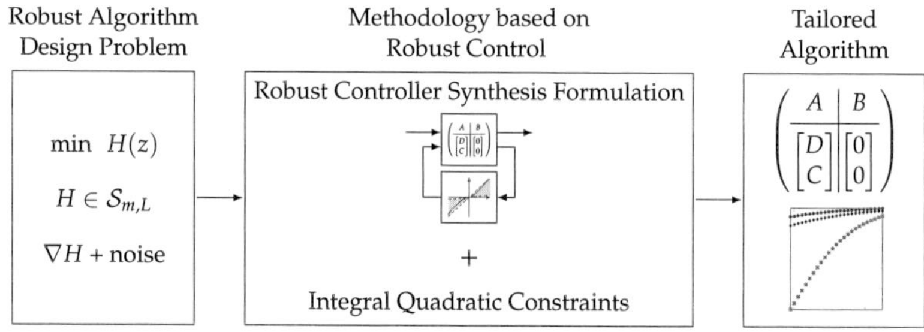

Figure 4.1. A schematic overview of the two classes of algorithm design problems considered in this thesis with the corresponding systems theoretic approach.

theory, in particular robust control theory. Taking this robust control viewpoint regarding the gradient of the objective function as the uncertain term (cf. Figure 4.1), we derived computationally tractable conditions in terms of LMIs allowing the analysis of existing and the design of novel optimization algorithms. In particular, we employed and suitably extended the framework of integral quadratic constraints (IQCs). We derived a class of IQCs valid for the considered class of strongly convex objective functions that extends the IQCs known from literature. Our results enable us to both analyze the convergence rates and robustness of existing algorithms and to design novel algorithms with specified guarantees. The proper embedding into the framework of integral quadratic constraints further allowed us to easily and systematically employ a variety of existing results. As an example, by making use of IQCs for structured uncertainties, we showed how structural properties of the objective function can be incorporated in the algorithm design.

There are several directions of possible future research. First, by the nature of our approach of embedding the two algorithm design problems in a control theoretic framework, both methodologies allow for systematic extensions. We already discussed some initial ideas in Section 2.5, Section 3.5, providing preliminary solutions or at least solution ideas. Besides these extensions, in Section 2.6, Section 3.6 we also proposed some open research questions that we regard as worth to consider. Concerning our vision of providing a systems theoretic approach to optimization, we think that it is worthwile to study how control theory can contribute further to the design of optimization algorithms.

Notation and Technical Background

A.1 Basic Notation

In the following we briefly introduce some basic notation; additional more specific notation is introduced in the subsequent sections Appendix A.2, Appendix A.3 and Appendix A.4.

Common Sets and Function Spaces. We let $\mathbb{N}$ denote the set of non-negative integers, $\mathbb{N}_{>0}$ the set of positive integers and denote by $\mathbb{Z}$ the set of all integers. Similarly, we denote by $\mathbb{R}^n$ the set of n-dimensional real vectors, by $\mathbb{R}^n_{\geq 0}$ those with non-negative entries and by $\mathbb{R}^n_{>0}$ those with positive entries. We further denote by $\mathbb{C}$ the set of all complex numbers and let $\mathbb{T} := \{z \in \mathbb{C} \mid |z| = 1\}$ denote the unit circle in the complex plane. We write $\mathcal{C}^p$, $p \in \mathbb{N}$, for the set of p-times continuously differentiable functions.

Common Matrices, Matrix Operations and Relations. For $n, m \in \mathbb{N}_{>0}$, we denote by $I_n \in \mathbb{R}^{n \times n}$ the $n \times n$ identity matrix, by $0_n \in \mathbb{R}^{n \times n}$ the $n \times n$ matrix of zeros and by $0_{n \times m} \in \mathbb{R}^{n \times m}$ the $n \times m$ matrix of zeros. Sometimes we omit the subscript if the dimensions are obvious. We denote the (i, j)th entry of a matrix $A \in \mathbb{R}^{n \times m}$ by a_{ij}, and sometimes denote A by $A = [a_{ij}]_{i=1,2,\ldots,n, j=1,2,\ldots,m}$. For a real vector $x = [x_1^\top, \ldots, x_n^\top]^\top$ and a finite set $S \subset \{1, \ldots, n\}$, we denote by $[x_i]_{i \in S}$ the stacked column vector of all x_i with $i \in S$ ordered by the index i. For example, for $n = 7$ and $S = \{6, 2, 3\}$, we have $[x_i]_{i \in S} = \begin{bmatrix} x_2^\top & x_3^\top & x_6^\top \end{bmatrix}^\top$. We let $\mathbf{1}$ denote the column vector with all entries being equal to one use e_i to denote the real vector with the ith entry equal to 1 and all other entries equal to 0, where, in both cases, we assume the dimension to be clear from the context. For a square matrix A we let $\mathrm{tr}(A)$ denote the trace of A. The rank of A is denoted by $\mathrm{rank}(A)$. For quadratic forms $x^* P x$, we often shortly write $(\star)^* P x$, where the superscript * denotes complex conjugation. We additionally replace blocks in symmetric matrices by the $\star$ symbol if they can be inferred from symmetry. We further denote by $\mathrm{blkdiag}(M_1, M_2, \ldots, M_k)$ the block diagonal matrix with blocks M_1, M_2, $\ldots$, M_k on its diagonal. If M_i, $i = 1, 2, \ldots, k$, are scalars, we also use $\mathrm{diag}(M_1, M_2, \ldots, M_k)$ instead to denote the respective diagonal matrix. For any two matrices A_1, A_2, we let $A_1 \otimes A_2$ denote the Kronecker product of A_1 with A_2. If A_1, A_2 are square, symmetric and of the same dimension, we write $A_1 \prec A_2$ ($A_1 \preceq A_2$) if $A_1 - A_2$ is negative definite (negative semi-definite) and $A_1 \succ A_2$ ($A_1 \succeq A_2$) if $A_1 - A_2$ is positive definite (positive semi-definite). In the same manner, we use the relations $<, \leq, >, \geq$ for elementwise comparison.

A.2 Basics of Graph Theory

We recall some basic notions on graph theory, and refer the reader to Biggs (1993) or other textbooks on graph theory for a more detailed introduction. A directed graph (or simply digraph) is an ordered pair $\mathcal{G} = (\mathcal{V}, \mathcal{E})$, where $\mathcal{V} = \{1, 2, \ldots, n\}$ is the set of nodes and $\mathcal{E} \subseteq \mathcal{V} \times \mathcal{V}$ is the set of edges, i.e., $(i, j) \in \mathcal{E}$ if there is an edge from node i to node j. An exemplary graph is depicted in Figure A.1, where the nodes are depicted by circles and the edges are indicated by arrows. In our setup, each node corresponds to an agent and the edges encode the communication topology amongst the network of agents, i.e., $(i, j) \in \mathcal{E}$ means that node i receives information from node j. In Figure A.1, for example, agent 1 has access to the state of agent 2. We say that node j is an *out-neighbor* of node i if there is an edge from node i to node j, i.e., $(i, j) \in \mathcal{E}$. For a graph $\mathcal{G}$, we then denote by

$$\mathcal{N}_{\mathcal{G}}(i) = \{i\} \cup \{j \in \mathcal{V} : (i, j) \in \mathcal{E}\} \tag{A.1}$$

the set of all out-neighbors of agent i as well as the agent itself. The *adjacency matrix* $\mathbf{A} = [\mathbf{a}_{ij}] \in \mathbb{R}^{n \times n}$ associated to $\mathcal{G}$ is defined as

$$\mathbf{a}_{ij} = \begin{cases} 1 & \text{if } i \neq j \text{ and } (i, j) \in \mathcal{E}, \\ 0 & \text{otherwise.} \end{cases} \tag{A.2}$$

We also define the *out-degree matrix* $\mathrm{D} = [\mathrm{d}_{ij}]$ associated to $\mathcal{G}$ as

$$\mathrm{d}_{ij} = \begin{cases} \sum_{k=1}^{n} \mathbf{a}_{ik} & \text{if } i = j \\ 0 & \text{otherwise.} \end{cases} \tag{A.3}$$

Finally, we call $G = \mathrm{D} - \mathbf{A} = [g_{ij}] \in \mathbb{R}^{n \times n}$ the *Laplacian* of $\mathcal{G}$. A digraph is said to be *undirected* if $(i, j) \in \mathcal{E}$ implies that $(j, i) \in \mathcal{E}$, or, equivalently, if $G = G^\top$. Further, a digraph $\mathcal{G}$ is called *weight-balanced* if $\mathbf{1}_n^T G = 0$. A (directed) path in $\mathcal{G}$ is a sequence of nodes connected by edges and we write $\mathfrak{p}_{i_1, i_r} = \langle i_1 \mid i_2 \mid \ldots \mid i_r \rangle$ for a path from node i_1 to node i_r. The path $\mathfrak{p}_{i_1, i_r}$ is said to be *simple* if it does not have repeated nodes, i.e., $i_j \neq i_k$ for any $j \neq k$, $j, k = 1, 2, \ldots, r$. We further denote by $\mathrm{head}(\mathfrak{p}_{i_1, i_r}) = i_1$ and $\mathrm{tail}(\mathfrak{p}_{i_1, i_r}) = i_r$ the head and the tail of a path $\mathfrak{p}_{i_1, i_r}$, respectively. We also let $\ell(\mathfrak{p}_{i_1, i_r}) = r - 1$ denote the length of the path. A digraph $\mathcal{G}$ is said to be *strongly connected* (or simply connected in case of undirected graphs) if there is a directed path between any two nodes. For a path $\mathfrak{p}_{i,j}$ from node i to node j we denote by $\mathrm{subpath}_{i\bullet}(\mathfrak{p}_{i,j})$ and $\mathrm{subpath}_{\bullet j}(\mathfrak{p}_{i,j})$ the set of all subpaths of $\mathfrak{p}_{i,j}$ (not including $\mathfrak{p}_{i,j}$ itself) which, respectively, start at i or end at j. Given a subpath $\mathfrak{q} \in \mathrm{subpath}_{i\bullet}(\mathfrak{p}_{i,j})$, we denote by $\mathfrak{q}^c$ the path in $\mathrm{subpath}_{\bullet j}(\mathfrak{p}_{i,j})$ whose composition with $\mathfrak{q}$ gives $\mathfrak{p}_{i,j}$.

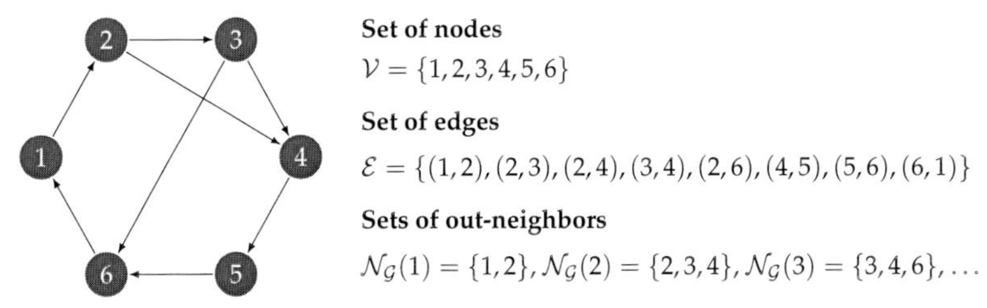

Figure A.1. An exemplary graph with six nodes and the corresponding node and edge set.

A.3 Lie Brackets and Formal Brackets

In the following, we give a very brief introduction to Lie brackets and formal brackets as required for the present work. For a more detailed treatment we refer the reader to standard textbooks on the subject, e.g., Bourbaki (1998).

Lie brackets of vector fields. Let $\mathbf{\Phi} = \{\phi_1, \phi_2, \ldots, \phi_M\}$ be a set of smooth vector fields $\phi_k : \mathbb{R}^n \to \mathbb{R}^n$, $k = 1, 2, \ldots, M$. For two vector field $\phi_i, \phi_j \in \mathbf{\Phi}$, we define the *Lie bracket* of ϕ_i, ϕ_j as

$$[\phi_i, \phi_j] : x \mapsto \frac{\partial \phi_2}{\partial x}(x)\phi_1(x) - \frac{\partial \phi_1}{\partial x}(x)\phi_2(x); \tag{A.4}$$

hence the Lie bracket of vector fields is a map that assigns to any two differentiable vector fields ϕ_i, ϕ_j another vector field $[\phi_i, \phi_j]$, which we call the Lie bracket. For convenience, we sometimes directly use the argument of the Lie bracket inside the bracket; e.g., we write $[x^2, \cos(x)]$ instead of $[\phi_1, \phi_2](x)$ with $\phi_1(x) = x^2, \phi_2(x) = \cos(x)$. Lie brackets of vector fields, and Lie brackets in general, have the following properties:

1. Bilinearity: For all $a, b \in \mathbb{R}$, $\phi_i, \phi_j, \phi_k \in \mathbf{\Phi}$ we have for all $x \in \mathbb{R}^n$

$$[a\phi_i + b\phi_j, \phi_k](x) = a[\phi_i, \phi_k](x) + b[\phi_j, \phi_k](x). \tag{A.5}$$

2. Skew-symmetry: For all $\phi_i, \phi_j \in \mathbf{\Phi}$ we have for all $x \in \mathbb{R}^n$

$$[\phi_i, \phi_j](x) = -[\phi_j, \phi_i](x). \tag{A.6}$$

3. Jacobi-identity: For all $\phi_i, \phi_j, \phi_k \in \mathbf{\Phi}$ we have for all $x \in \mathbb{R}^n$

$$\big[\phi_i, [\phi_j, \phi_k]\big](x) = -\big[\phi_j, [\phi_k, \phi_i]\big](x) - \big[\phi_k, [\phi_i, \phi_j]\big](x). \tag{A.7}$$

We denote by $\mathcal{LBr}(\mathbf{\Phi})$ the set of Lie brackets generated by $\mathbf{\Phi}$, i.e., the set of vector fields which contains $\mathbf{\Phi}$ and has the property that, if $B_1, B_2 \in \mathcal{LBr}(\mathbf{\Phi})$, then also $[B_1, B_2] \in \mathcal{LBr}(\mathbf{\Phi})$.

Formal brackets. Let $\mathbf{X} = \{X_1, X_2, \ldots, X_M\}$ be a finite set of M non-commuting objects, the so-called *indeterminates*. We denote by $\mathcal{FB}r(\mathbf{X})$ the set of formal brackets constructed from $\mathbf{X}$, where a formal bracket is a word which is constructed from the alphabet consisting of the symbols X_k, $k = 1, 2, \ldots, M$, in $\mathbf{X}$ as well as the brackets [and] and the comma ,. More precisely, the set of formal brackets $\mathcal{FB}r(\mathbf{X})$ is defined as the smallest set of words built from that alphabet which contains all elements of $\mathbf{X}$ and has the property that, for all $B_1, B_2 \in \mathcal{FB}r(\mathbf{X})$, the word $[B_1, B_2]$ is an element of $\mathcal{FB}r(\mathbf{X})$.

Relation between Lie brackets and formal brackets. The two concepts of Lie brackets and formal brackets are closely related. Let $\mu : \mathcal{FB}r(\mathbf{X}) \to \mathcal{LB}r(\mathbf{X})$ denote the mapping that maps formal brackets to Lie brackets and let $\mathrm{Ev} : \mathcal{LB}r(\mathbf{X}) \to \mathcal{LB}r(\mathbf{\Phi})$ denote the mapping that maps Lie brackets of indeterminates to Lie brackets of vector fields, simply by replacing the indeterminates by the corresponding vector fields. For example, it is

$$\mathrm{Ev}\big([X_1, [X_1, X_2]]\big) = \big[\phi_1, [\phi_1, \phi_2]\big]. \tag{A.8}$$

In rough words, a formal bracket can be interpreted as a string representation of a Lie bracket. However, this string representation is in general not unique. As an example, we distinguish between the two formal brackets $\big[X_1, [X_1, X_2]\big]$ and $\big[[X_2, X_1], X_1\big]$, but these brackets are equal as Lie brackets by the skew-symmetry property (A.6), thus μ is not injective. Similarly, the map Ev is not injective, too, since two Lie brackets of vector fields might yield the same vector field while they are different as Lie brackets of indeterminates. In a nutshell, this is the main reason why formal brackets are formally required in the present work; more specifically several operators, which we will introduce in the following, cannot be defined for Lie brackets but need to be defined on formal brackets since the following chain is not reversible:

$$\mathcal{FB}r(\mathbf{X}) \xrightarrow{\quad \mu \quad} \mathcal{LB}r(\mathbf{X}) \xrightarrow{\quad \mathrm{Ev} \quad} \mathcal{LB}r(\mathbf{\Phi}).$$

Operators on formal brackets and their Lie bracket interpretation. We next define the required operators on formal brackets; since we are also abusing them in a formally incorrect manner on Lie brackets, we additionally explain how they are to be interpreted. Let $\mathbf{\Phi}$, $\mathbf{X}$, $\mathcal{FB}r(\mathbf{X})$, $\mathcal{LB}r(\mathbf{\Phi})$ be defined as before. For any $B = [B_1, B_2] \in \mathcal{FB}r(\mathbf{X})$, we define the left and right factor left, right : $\mathcal{FB}r(\mathbf{X}) \to \mathcal{FB}r(\mathbf{X})$ as

$$\mathrm{left}([B_1, B_2]) = B_1, \quad \mathrm{right}([B_1, B_2]) = B_2. \tag{A.9}$$

For $B \in \mathcal{FB}r(\mathbf{X})$, we then let $\delta : \mathcal{FB}r(\mathbf{X}) \to \mathbb{N}_{>0}$ denote the degree defined as

$$\delta(B) = \begin{cases} 1 & \text{if } B \in \mathbf{X} \\ \delta\big(\mathrm{left}(B)\big) + \delta\big(\mathrm{right}(B)\big) & \text{otherwise.} \end{cases} \tag{A.10}$$

Similarly, we let $\delta_k(B)$, $k \in \{1,2,\ldots,M\}$ denote the degree of the kth indeterminate in B, i.e.,

$$\delta_k(B) = \begin{cases} 1 & \text{if } B = X_k \\ 0 & \text{if } B \in \mathbf{X} \setminus \{X_k\} \\ \delta_k(\text{left}(B)) + \delta_k(\text{right}(B)) & \text{otherwise.} \end{cases} \tag{A.11}$$

In rough words, $\delta(B)$ is the total number of indeterminates in B and $\delta_k(B)$ is the number of times the indeterminate X_k appears in B. Utilizing these operators, we then define a so-called *P. Hall set* as a subset of the set of formal brackets as follows:

Definition 13 (P. Hall set). Let $\mathbf{X} = \{X_1, X_2, \ldots, X_M\}$ be a set of indeterminates. A *P. Hall set* $\mathcal{PH}(\mathbf{X}) = (\mathbb{B}, \prec)$ is a set $\mathbb{B}$ of brackets equipped with a total ordering $\prec$ that fulfills the following properties:

[PH1] Every X_k, $k = 1,2,\ldots,M$, is in $\mathbb{B}$.

[PH2] $X_k \prec X_j$ if and only if $k < j$.

[PH3] If $B_1, B_2 \in \mathbb{B}$ and $\delta(B_1) < \delta(B_2)$, then $B_1 \prec B_2$.

[PH4] Each $B = [B_1, B_2] \in \mathbb{B}$ if and only if

[PH4.a] $B_1, B_2 \in \mathbb{B}$ and $B_1 \prec B_2$

[PH4.b] either $\delta(B_2) = 1$ or $B_2 = [B_3, B_4]$ for some B_3, B_4 such that $B_3 \preceq B_1$. •

By construction, a P. Hall set $\mathcal{PH}(\mathbf{X})$ has the property that the restriction of μ to $\mathcal{PH}(\mathbf{X})$ is a bijective map. It is clear that the latter operators and P. Hall sets cannot be defined properly for Lie brackets of vector fields since a Lie bracket of vector fields does not yield a unique decomposition in its left and right factor. Still, if the reader accepts to interpret a Lie bracket of vector fields in its explicit representation as a Lie bracket bracket of indeterminates, we think that it is worth to avoid the overhead introduced by being formally correct in favor of a clearer presentation.

A.4 Sequence Spaces and (Linear) Operators

Sequence spaces. In the following we introduce some standard sequence spaces. We limit ourselves to one-sided sequences here, and denote a sequence $q : \mathbb{N} \to \mathbb{R}^p$, $q : k \mapsto q_k$, by $(q_k)_{k\in\mathbb{N}} = (q_0, q_1, q_2, \ldots)$, where $q_k \in \mathbb{R}^p$, $k \in \mathbb{N}$. We then let $\ell_e^p = \{(q_k)_{k\in\mathbb{N}} \mid q_k \in \mathbb{R}^p\}$ denote the subspace of all one-sided sequences and denote by $\ell_f^p = \{q \in \ell_e^p \mid \exists T \in \mathbb{N} \text{ s.t. } q_k = 0 \text{ for all } k > T\}$ the set of all finitely supported sequences therein. We further introduce the space of all square-summable sequences $\ell_2^p \subset \ell_e^p$ defined by

$$\ell_2^p = \{q \in \ell_e^p \mid \sum_{k=0}^{\infty} \|q_k\|^2 < \infty\}. \tag{A.12}$$

We sometimes omit the dimension of the signals and write

$$\ell_e := \bigcup_{i\in\mathbb{N}_{>0}} \ell_e^i, \quad \ell_2 := \bigcup_{i\in\mathbb{N}_{>0}} \ell_2^i, \quad \ell_f := \bigcup_{i\in\mathbb{N}_{>0}} \ell_f^i \tag{A.13}$$

for the collection of all sequences of the respective type. We denote the standard inner product by $\langle u, y\rangle = \sum_{k=0}^{\infty} u_k^\top y_k$, where $u, y \in \ell_2^p$ and the superscript $^\top$ denotes transposition. We further let $\|y\|_{\ell_2} = \sqrt{\langle y, y\rangle}$ denote the corresponding induced norm.

Operators. Operators are mappings $\phi : \ell_f^p \to \ell_e^q$, $p, q \in \mathbb{N}_{>0}$, i.e, an operator ϕ maps a sequence $u \in \ell_f^p$ to a sequence $y = \phi(u) \in \ell_e^q$. We say that an operator $\phi : \mathcal{U} \to \mathcal{V}$, $\mathcal{U}, \mathcal{V} \subseteq \ell_2^p$, is bounded (on $\mathcal{U}$) if there exists $\beta \geq 0$ such that $\|\phi(y)\|_{\ell_2} \leq \beta\|y\|_{\ell_2}$ for all $y \in \mathcal{U}$. If $\mathcal{U} = \mathcal{V} = \ell_2^p$, we call the infimal β that fulfills this inequality the ℓ_2-gain of ϕ. We further let $\mathcal{L}(\mathcal{U}, \mathcal{V})$ denote the set of operators mapping $\mathcal{U}$ to $\mathcal{V}$ that are bounded on $\mathcal{U}$. For two operators $\phi_1 : \ell_f^p \to \ell_e^q$, $\phi_2 : \ell_f^q \to \ell_e^r$, we let $\phi_2 \circ \phi_1$ denote the composition, i.e., $\phi_2 \circ \phi_1$ maps any input $u \in \ell_e^p$ to $y_2 = \phi_2(\phi_1(u)) \in \ell_e^r$. For linear operators ϕ_1, we often omit the brackets and simply write $y_1 = \phi_1 u$ instead of $y_1 = \phi_1(u)$. We further let $\mathrm{id} : \ell_e^p \to \ell_e^p$, $p \in \mathbb{N}_{>0}$, denote the identity operator.

z-Transforms, Transfer Functions and Toeplitz Operators. If it exists, i.e., if it is well-defined, we denote by $\widehat{q} = \mathcal{Z}(q)$ the one-sided z-transform of a signal $q \in \ell_2^p$ which is defined by

$$\widehat{q}(z) = \mathcal{Z}(q)(z) = \sum_{k=0}^{\infty} q_k z^{-k}, \tag{A.14}$$

where $\widehat{q} : \mathbb{U} \to \mathbb{C}^p$ for some set $\mathbb{U} \subseteq \mathbb{C}$. We denote by $\mathcal{RH}_\infty^{n\times m}$ the set of all real-rational and proper transfer matrices of dimension $n \times m$ having all poles in the open unit disk. Similarly, we denote by $\mathcal{RL}_\infty^{n\times m}$ the set of all real-rational and proper transfer matrices of dimension $n \times m$ having no poles on the unit circle. For a square transfer matrix G we write

$$G \overset{\mathbb{S}}{\prec} 0 \tag{A.15}$$

if the matrix $G(z) + G^*(z)$ is negative definite for all $z \in \mathbb{S} \subseteq \mathbb{C}$. For any transfer matrix $G \in \mathcal{RL}_\infty^{n\times m}$, we let $\phi_G : \ell_2^m \to \ell_2^n$ denote the corresponding linear Toeplitz operator in the time domain, i.e., ϕ_G fulfills

$$\mathcal{Z}(\phi_G u) = G\widehat{u} \tag{A.16}$$

for any $u \in \ell_2^m$. Note that when ϕ_G is interpreted as an infinite-dimensional matrix acting on u, then ϕ_G has a block Toeplitz structure, where the blocks are of size $n \times m$. For a proper transfer matrix $G(z) = C(zI - A)^{-1}B + D$ we write $G \sim (A, B, C, D)$ or

$$G \sim \left(\begin{array}{c|c} A & B \\ \hline C & D \end{array}\right) \tag{A.17}$$

to indicate that G admits the state-space realization (A, B, C, D).

A.5 Convex Functions and Memoryless Monotone Operators

For the sake of completeness, we repeat some important definitions and standard results concerning convex functions. We refer the reader to standard textbooks on convex analysis and optimization for further information, see, e.g., Boyd and Vandenberghe (2004); Polyak (1987).

Definition 14 ((Strictly/Strongly) Convex Functions). A function $H : \mathbb{R}^p \to \mathbb{R}$ is said to be

(a) *convex*, if, for any $x, y \in \mathbb{R}^p$, $\lambda \in [0,1]$,

$$H\big(\lambda x + (1-\lambda)y\big) \leq \lambda H(x) + (1-\lambda)H(y); \tag{A.18}$$

(b) *strictly convex*, if, for any $x, y \in \mathbb{R}^p$, $x \neq y$, $\lambda \in [0,1]$,

$$H\big(\lambda x + (1-\lambda)y\big) < \lambda H(x) + (1-\lambda)H(y); \tag{A.19}$$

(c) *m-strongly convex*, $m > 0$, if, for any $x, y \in \mathbb{R}^p$, $\lambda \in [0,1]$,

$$H\big(\lambda x + (1-\lambda)y\big) \leq \lambda H(x) + (1-\lambda)H(y) - \tfrac{1}{2}m\lambda(1-\lambda)\|x-y\|^2. \tag{A.20}$$

•

For differentiable functions, the following result is well-known.

Lemma 13. The function $H : \mathbb{R}^p \to \mathbb{R}$, $H \in \mathcal{C}^1$, is

(a) *convex*, if and only if, for any $x, y \in \mathbb{R}^p$,

$$H(x) \geq H(y) + \langle \nabla H(y), x-y \rangle; \tag{A.21}$$

(b) *strictly convex*, if and only if, for any $x, y \in \mathbb{R}^p$, $x \neq y$,

$$H(x) > H(y) + \langle \nabla H(y), x-y \rangle; \tag{A.22}$$

(c) *m-strongly convex*, if and only if, for any $x, y \in \mathbb{R}^p$,

$$H(x) \geq H(y) + \langle \nabla H(y), x-y \rangle + \tfrac{1}{2}m\|x-y\|^2. \tag{A.23}$$

•

Many equivalent characterizations exist in the literature; for example, it is well-known (Nesterov, 2004, Theorem 2.1.10) that H is m-strongly convex if and only if, for any $x, y \in \mathbb{R}^p$,

$$\big(\nabla H(x) - \nabla H(y)\big)^\top (x-y) \geq m\|x-y\|^2. \tag{A.24}$$

We often consider the following class of (strongly) convex objective functions with Lipschitz continuous gradient:

Definition 15 (Class $\mathcal{S}_{m,L}$). A function $H : \mathbb{R}^p \to \mathbb{R}$ is said to belong to the class $\mathcal{S}_{m,L}$, $L \geq m \geq 0$, if $H \in \mathcal{C}^2$ and

$$\left(\nabla H(x) - \nabla H(y)\right)^\top (x - y) \geq m\|x - y\|^2 \tag{A.25a}$$

$$\|\nabla H(x) - \nabla H(y)\| \leq L\|x - y\| \tag{A.25b}$$

for all $x, y \in \mathbb{R}^p$. •

In view of a lighter notation, we do not explicitly specify the dimension of the domain but tacitly assume it to be clear from the context. Note that if $m > 0$, then, by (A.24), H is strongly convex and H has a unique minimum. The so defined class of functions is well-known in the optimization literature, see, e.g., Nesterov (2004) for further properties and details. In particular, the following result (Nesterov, 2004, Theorem 2.1.11) combining (A.24) and (A.25b) is important.

Lemma 14. Let $H : \mathbb{R}^p \to \mathbb{R}$, $H \in \mathcal{S}_{m,L}$ for some $L \geq m > 0$. Then

$$\left(\nabla H(x+y) - \nabla H(y)\right)^\top y \geq \frac{mL}{m+L}\|y\|^2 + \frac{1}{m+L}\|\nabla H(x+y) - \nabla H(y)\|^2 \tag{A.26}$$

for all $x, y \in \mathbb{R}^p$ •

An equivalent characterization of the class $\mathcal{S}_{m,L}$ can be given in terms of the eigenvalues of the Hessian (Nesterov, 2004, Theorem 2.1.6, 2.1.11).

Lemma 15. A function $H : \mathbb{R}^p \to \mathbb{R}$, $H \in \mathcal{C}^2$, belongs to the class $\mathcal{S}_{m,L}$, $L \geq m \geq 0$, if and only if it is

$$mI \preceq \nabla^2 H(x) \preceq LI \tag{A.27}$$

for all $x \in \mathbb{R}^p$. •

We continue with the definitions of monotone and slope-restricted memoryless operators, two concepts closely related to convex functions and the class $\mathcal{S}_{m,L}$. To this end, we first define monotone functions as follows.

Definition 16 (Monotone/Slope-Restricted/Sector Bounded Function). A continuously differentiable function $f : \mathbb{R}^p \to \mathbb{R}^p$ is said to be *monotone* if (1) it is conservative, i.e., there exists $F : \mathbb{R}^p \to \mathbb{R}$ such that $\nabla F(x) = f(x)$ for all $x \in \mathbb{R}^p$ and (2) it fulfills

$$\langle f(x) - f(y), x - y\rangle \geq 0 \tag{A.28}$$

for all $x, y \in \mathbb{R}^p$. We say that f is *slope-restricted* in (α, β), $\beta \geq \alpha \geq 0$, if it is monotone and

$$\alpha\|x - y\|^2 \leq \langle f(x) - f(y), x - y\rangle \leq \beta\|x - y\|^2 \tag{A.29}$$

for all $x, y \in \mathbb{R}^p$; we say that f is *sector bounded* in (α, β), $\beta \geq \alpha$, if $f(0) = 0$ and

$$\langle f(x) - \alpha x, \beta x - f(x)\rangle \geq 0 \tag{A.30}$$

for all $x \in \mathbb{R}^p$. •

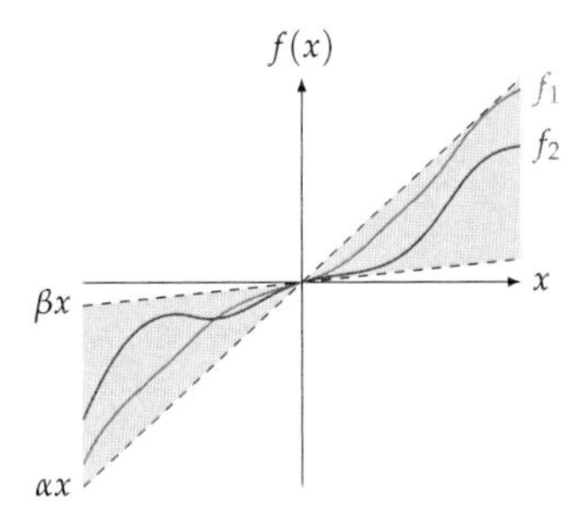

Figure A.2. An illustration of sector bounded and slope-restricted functions of one variable. The function f_1 is sector bounded and slope-restricted while f_2 is sector bounded but not slope-restricted.

The requirement (1) that f is conservative is not standard in the definition of monotone functions; we note that in case of $p = 1$ it is also not restrictive. Clearly, a function is monotone if it is the gradient of some convex function; it is slope-restricted in (α, β) if it is from the class $\mathcal{S}^1_{\alpha,\beta}$. We note that a slope-restricted function in (α, β) is sector bounded in (α, β); the converse, however, is not true in general, see Figure A.2 for an illustration. Monotone (respectively, slope-restricted or sector bounded) operators $\varphi : \ell_e^p \to \ell_e^p$ are then defined as operators of the form

$$\big(\varphi(y)\big)_k = f(y_k), \tag{A.31}$$

where $y = [y_0, y_1, \ldots] \in \ell_e^p$, $k \in \mathbb{N}$, and $f : \mathbb{R}^p \to \mathbb{R}^p$ is a monotone (respectively, slope-restricted or sector bounded) function.

A.6 Some Facts Involving Linear Matrix Inequalities (LMIs)

For the sake of completeness, we repeat some standard facts involving linear matrix inequalities; these results can be found in standard textbooks, e.g., Boyd, El Ghaoui, Feron, and Balakrishnan (1994).

Lemma 16 (Discrete-Time KYP-Lemma). Let $G \in \mathcal{RH}_\infty^{n\times m}$, $n, m \in \mathbb{N}_{>0}$, be a real-rational and proper transfer matrix and let (A, B, C, D) denote a state-space representation of G. Suppose that A has no eigenvalues on the unit circle and let $M \in \mathbb{R}^{n\times n}$ be a real symmetric matrix. Then the following frequency domain inequality

$$G(z)^* M G(z) \overset{\mathbb{T}}{\prec} 0 \tag{A.32}$$

holds if and only if there exists a $P = P^\top$ of appropriate dimension such that

$$\begin{bmatrix} A & B \\ I & 0 \\ C & D \end{bmatrix}^\top \begin{bmatrix} P & 0 & 0 \\ 0 & -P & 0 \\ 0 & 0 & M \end{bmatrix} \begin{bmatrix} A & B \\ I & 0 \\ C & D \end{bmatrix} \prec 0. \tag{A.33}$$

•

Lemma 17 (Schur Complements). Let $X \in \mathbb{R}^{n\times n}$, $n \geq 2$, be a symmetric matrix decomposed as

$$X = \begin{bmatrix} X_1 & X_2 \\ X_2^\top & X_3 \end{bmatrix}. \tag{A.34}$$

Then the following statements are equivalent:

1. $X \succ 0$.

2. $X_1 \succ 0$ and $X_3 - X_2^\top X_1^{-1} X_2 \succ 0$.

3. $X_3 \succ 0$ and $X_1 - X_2 X_3^{-1} X_2^\top \succ 0$. •

Lemma 18 (S-Procedure for Quadratic Forms). Let $T_0, T_1 \mathbb{R}^{n\times n}$, $n \in \mathbb{N}_{>0}$, be two symmetric matrices. Then $x^\top T_0 x > 0$ for all x with $x^\top T_1 x \geq 0$ if there exists a $\lambda \geq 0$ such that $T_0 - \lambda T_1 \succ 0$. Moreover, under the assumption that there exists an $x_0 \in \mathbb{R}^n$ such that $x_0^\top T_0 x_0 > 0$, this condition is not only sufficient but also necessary, i.e. it is $x^\top T_0 x > 0$ for all x with $x^\top T_1 x \geq 0$ if and only if there exists a $\lambda \geq 0$ such that $T_0 - \lambda T_1 \succ 0$. •

B

Proofs

In this chapter, we collect all proofs of the results derived in Chapter 2 and Chapter 3.

B.1 Proofs of Chapter 2

B.1.1 Proof of Theorem 1

The proof follows a similar argument as the one in Dürr (2015, Theorem 5.1.3). First, using (2.19c), we have

$$\lambda_i(t) = \exp\big(\int_0^t c_i(z(\tau))\mathrm{d}\tau\big)\lambda_i(0), \tag{B.1}$$

for all $i \in \{1, 2, \ldots, n_{\text{ineq}}\}$; hence, $\lambda_i(0) > 0$ implies that $\lambda_i(t) > 0$, for all $t \geq 0$, and consequently, the set $\mathcal{R}(\mathcal{M})$ is positively invariant w.r.t. (2.19). Let $(z^\star, \nu^\star, \lambda^\star)$ be an arbitrary point in $\mathcal{M}$. Consider the candidate Lyapunov function $V : \mathbb{R}^p \times \mathbb{R}^{n_{\text{eq}}} \times \mathbb{R}_{\geq 0}^{n_{\text{ineq}}} \to \mathbb{R}_{\geq 0}$ defined as

$$V(z, \nu, \lambda) = \tfrac{1}{2}\|z - z^\star\|^2 + \tfrac{1}{2}\|\nu - \nu^\star\|^2 + \sum_{i=1}^{n}(\lambda_i - \lambda_i^\star) - \sum_{i:\lambda_i^\star \neq 0}^{n_{\text{ineq}}} \lambda_i^\star \ln(\tfrac{\lambda_i}{\lambda_i^\star}). \tag{B.2}$$

We first observe that V is positive definite with respect to $(z^\star, \nu^\star, \lambda^\star)$ on $\mathcal{R}(\mathcal{M})$, and that all the level sets are compact. To see this, note that according to (Bregman, 1967, p. 207, eq. (1.5)), the function $D : \mathbb{R}_{\geq 0}^{n_{\text{ineq}}} \times \mathbb{R}_{\geq 0}^{n_{\text{ineq}}} \to \mathbb{R}$ defined as

$$D(\lambda^\star, \lambda) = \sum_{i=1}^{n_{\text{ineq}}} \big(\lambda_i - \lambda_i^\star + \lambda_i^\star(\ln(\lambda_i^\star) - \ln(\lambda_i))\big) = \sum_{i=1}^{n_{\text{ineq}}}(\lambda_i - \lambda_i^\star) - \sum_{i:\lambda_i^\star \neq 0}^{n_{\text{ineq}}} \lambda_i^\star \ln(\tfrac{\lambda_i}{\lambda_i^\star}) \tag{B.3}$$

is non-negative for all $(\lambda^\star, \lambda) \in \mathbb{R}_{\geq 0}^{n_{\text{ineq}}} \times \mathbb{R}_{\geq 0}^{n_{\text{ineq}}}$ and zero if and only if $\lambda = \lambda^\star$ (Bregman, 1967, Condition I.) and its level sets are compact (Bregman, 1967, Condition V.). Thus, with V additionally being quadratic in z and ν, positive definiteness and compactness of all level sets follows and hence V is uniformly unbounded on $\mathcal{R}(\mathcal{M})$. The derivative of V along the trajectories of (2.19) is then given by

$$\dot{V}(z,\nu,\lambda) = -(z-z^\star)^\top\left(\nabla H(z) + \tfrac{da}{dz}(z)^\top \nu + \tfrac{dc}{dz}(z)^\top\lambda\right) + (\nu-\nu^\star)^\top a(z) \\ + \lambda^\top c(z) - {\lambda^\star}^\top c(z). \tag{B.4}$$

With a being affine according to **[A2]**, we have that[1]

$$(z-z^\star)^\top \tfrac{da}{dz}(z)^\top \nu = \nu^\top\left(a(z) - a(z^\star)\right). \tag{B.5}$$

Additionally, by the convexity assumptions **[A1]**, **[A2]**, for all $z \neq z^\star$ and $\lambda \in \mathbb{R}^{n_{\text{ineq}}}_{\geq 0}$ it is

$$-(z-z^\star)^\top \nabla H(z) < H(z^\star) - H(z) \tag{B.6}$$

$$-(z-z^\star)^\top \tfrac{dc}{dz}(z)^\top\lambda \leq \left(c(z^\star) - c(z)\right)^\top \lambda. \tag{B.7}$$

Thus, for all $z \neq z^\star$, $\lambda \in \mathbb{R}^{n_{\text{ineq}}}_{\geq 0}$, we infer

$$\dot{V}(z,\nu,\lambda) < H(z^\star) - H(z) - \nu^\top\left(a(z) - a(z^\star)\right) + \left(c(z^\star) - c(z)\right)^\top\lambda + (\nu-\nu^\star)^\top a(z) \\ + (\lambda^\top - \lambda^\star)^\top c(z). \tag{B.8}$$

Adding $0 = H(z) - H(z)$, and using the definition of the Lagrangian we have

$$\dot{V}(z,\nu,\lambda) < L(z^\star,\nu,\lambda) - L(z,\nu^\star,\lambda^\star). \tag{B.9}$$

Due to the saddle point property (2.17) the derivative of V along the trajectories of (2.19) is strictly negative, for all (z,ν,λ) except for $(z,\nu,\lambda) \in \mathcal{M}$; hence $(z^\star,\nu^\star,\lambda^\star)$ is stable according to (Dürr, 2015, Theorem 2.2.2). This procedure can be repeated for any point $(z^\star,\nu^\star,\lambda^\star) \in \mathcal{M}$, hence $\mathcal{M}$ is stable. We further note that the set $\mathcal{M}$ is compact due to Assumption 2. The same argument as the one in the proof of (Dürr, 2015, Theorem 5.1.3) then yields that the set of saddle points is $\mathcal{R}(\mathcal{M})$-globally uniformly asymptotically stable for (2.19).

B.1.2 Proof of Lemma 1

We prove this result by induction. For $l = r-1$, we have by (2.38), (2.40)

$$R_\psi(\mathfrak{p}_{r-1,r})(x) = h_{r-1,j_{r-1}}(x) = e_{j_{r-1}} F_{j_{r-1}}(\mathbf{x}_{r-1}, \mathbf{x}_r) \tag{B.10}$$

and the result follows immediately. Now suppose that the claim holds for all l with $r-1 \geq l \geq \bar{l}$, where $\bar{l} \in \{2,3,\ldots,r-1\}$. By (2.41), $R_\psi(\mathfrak{p}_{\bar{l}-1,r})(x) = \left[R_\psi(\mathfrak{q}^c), R_\psi(\mathfrak{q})\right](x)$, and any

[1] This is most easily be seen noting that $a(z) = Az - b$ for some $A \in \mathbb{R}^{n_{\text{eq}} \times n_{\text{eq}}}, b \in \mathbb{R}^{n_{\text{eq}}}$.

subpath $\mathfrak{q} \in \text{subpath}_{i_\bullet}(\mathfrak{p}_{\bar{l}-1,r})$ takes the form $\mathfrak{q} = \mathfrak{p}_{\bar{l}-1,t}$, $t \in \{\bar{l}, \bar{l}+1, \ldots, r-1\}$. Using the induction hypothesis, we then have

$$\begin{aligned} & R_\psi(\mathfrak{p}_{\bar{l}-1,r})(x) \\ &= \left[R_\psi(\mathfrak{p}_{t,r}), R_\psi(\mathfrak{p}_{\bar{l}-1,t})\right](x) \\ &= \left[e_{j_t}\tilde{F}_{t,r}, e_{j_{\bar{l}-1}}\tilde{F}_{\bar{l}-1,t}\right](x) \\ &= e_{j_{\bar{l}-1}}\Big(\sum_{k=\bar{l}-1}^{t} \frac{\partial}{\partial \mathbf{x}_k}\tilde{F}_{\bar{l}-1,t}(x)\Big)e_{j_t}\tilde{F}_{t,r}(x) - e_{j_t}\Big(\sum_{k=t}^{r} \frac{\partial}{\partial \mathbf{x}_k}\tilde{F}_{t,r}(x)\Big)e_{j_{\bar{l}-1}}\tilde{F}_{\bar{l}-1,t}(x). \end{aligned} \tag{B.11}$$

By (2.43), $\tilde{F}_{l,r}$ is a function of $\mathbf{x}_l, \mathbf{x}_{l+1}, \ldots, \mathbf{x}_r$ only. Hence, since $\mathfrak{p}_{1,r}$ is a simple path, $\tilde{F}_{t,r}$ is independent of $\mathbf{x}_{\bar{l}-1}$ and the second term vanishes. We then obtain

$$\begin{aligned} R_\psi(\mathfrak{p}_{\bar{l}-1,r})(x) &= e_{j_{\bar{l}-1}}\frac{\partial F_{j_{t-1}}}{\partial x_{j_t}}(\mathbf{x}_{r-1}, \mathbf{x}_r)\prod_{k=\bar{l}-1}^{t-2}\frac{\partial F_{j_k}}{\partial x_{j_{k+1}}}(\mathbf{x}_{i_k}, \mathbf{x}_{i_{k+1}})\tilde{F}_{t,r}(x) \\ &= e_{j_{\bar{l}-1}}\tilde{F}_{\bar{l}-1,r}(x), \end{aligned} \tag{B.12}$$

thus concluding the proof.

B.1.3 Proof of Lemma 2

For V not identically zero and W non-constant, it is clear that the structure (2.55) is not only sufficient but also necessary for $R_\psi(\mathfrak{p}_{i_1,i_r})(x)$ to be independent of $\mathbf{x}_{i_k}$, $k = 2, 3, \ldots, r-1$. Suppose now there exists some map $\psi : \langle i_k \mid i_{k+1}\rangle \mapsto j_k$, $j_k \in \mathcal{I}(i_k)$, and a set of bounded functions F_{j_k} fulfilling (2.55) such that (2.58) holds. Since $\mathcal{I}(i_k)$ and $\mathcal{I}(i_{k+1})$ are disjunct for any $k = 1, 2, \ldots, r-1$, by (2.55) this implies that V, W_{j_k}, $k = 1, 2, \ldots, r-1$, must be bounded. From (2.55b) we then infer that, for any $k = 2, \ldots, r-1$, there exists some constant $\beta > 0$ such that

$$\Big|\Big(\frac{\partial W_{j_{k-1}}}{\partial x_{j_k}}(\mathbf{x}_{i_k})\Big)^{-1}\Big| = \Big|\frac{F_{j_k}(\mathbf{x}_{i_k}, \mathbf{x}_{i_{k+1}})}{W_{j_k}(\mathbf{x}_{i_{k+1}})}\Big| \geq \beta \tag{B.13}$$

for all $x \in \mathbb{R}^N$. Hence $W_{j_{k-1}}(\mathbf{x}_{i_k})$ is strictly monotone in x_{j_k} which contradicts the boundedness assumption, thus concluding the proof.

B.1.4 Proof of Lemma 3

As an intermediate result, we first prove by induction that

$$\Big[\Big[\cdots\big[[\psi_{k_1}, \psi_{k_2}], \psi_{k_3}\big], \ldots\Big], \psi_{k_{M-1}}\Big](x) = e_j \prod_{t=1}^{M-1} W(\mathbf{x}_{k_t}) \tag{B.14}$$

for any $x \in \mathbb{R}^N$. For $M = 2$, it is clear that (B.14) holds. We impose the induction hypothesis that (B.14) holds for any $M \leq \bar{M}$, with $\bar{M} \geq 2$. For $M = \bar{M} + 1$ we then infer

$$
\begin{aligned}
&\Big[\Big[\cdots[[\psi_{k_1}, \psi_{k_2}], \psi_{k_3}], \dots\Big], \psi_{k_{\bar{M}}}\Big](x) \\
&= [e_j \prod_{t=1}^{\bar{M}-1} W_{k_t}, e_j x_j W_{k_{\bar{M}}}](x) \\
&= e_j (e_j^\top W_{k_{\bar{M}}}(\mathbf{x}_{k_{\bar{M}}}) + x_j \tfrac{\mathrm{d}W_{k_{\bar{M}}}(\mathbf{x}_{k_{\bar{M}}})}{\mathrm{d}x}) e_j \prod_{t=1}^{\bar{M}-1} W_{k_t} - \underbrace{e_j \tfrac{\mathrm{d}}{\mathrm{d}x}(\prod_{t=1}^{\bar{M}-1} W_{k_t}(\mathbf{x}_{k_t})) e_j x_j W_{k_{\bar{M}}}}_{=0} \\
&= e_j \prod_{t=1}^{\bar{M}} W(\mathbf{x}_{k_t});
\end{aligned}
\tag{B.15}
$$

thus proving (B.14). Using (B.14), we directly obtain with a slight abuse of notation

$$
\begin{aligned}
&\Big[\Big[\cdots[[\psi_{k_1}, \psi_{k_2}], \psi_{k_3}], \dots\Big], \psi_{k_M}\Big](x) \\
&= [e_j \prod_{t=1}^{M-1} W(\mathbf{x}_{k_t}), e_j W_{k_M}(\mathbf{x}_{k_M}) \int V(\mathbf{x}_i) \mathrm{d}x_j] \\
&= e_j V(\mathbf{x}_i) \prod_{t=1}^{M} W(\mathbf{x}_k),
\end{aligned}
\tag{B.16}
$$

thus concluding the proof.

B.1.5 Proof of Lemma 5

We first observe first that (2.102) is the same as (2.40), (2.41) with a special choice of the subpath as well as an additional projection with the property $\mathrm{proj}_{\mathbb{P}}(B)(x) = B(x)$ for all $x \in \mathbb{R}^N$. Hence, it immediately follows that $\tilde{R}_\psi(\mathfrak{p}_{i_1,i_r})(x) = R_\psi(\mathfrak{p}_{i_1,i_r})(x)$. In the same manner, we also have that

$$
\delta(\tilde{R}_\psi(\mathfrak{p}_{i_1,i_r})) = \delta(R_\psi(\mathfrak{p}_{i_1,i_r})) = \ell(\mathfrak{p}_{i_1,i_r}). \tag{B.17}
$$

We show the second part by induction. First observe that for paths $\mathfrak{p}_{i_1,i_r}$ with $\ell(\mathfrak{p}_{i_1,i_r}) = 1$ it is clear that $\tilde{R}_\psi(\mathfrak{p}_{i_1,i_r}) \in \mathbb{P}$ since $R_\psi(\mathfrak{p}_{i_1,i_r})$ is an admissible vector field by Lemma 1 and all admissible vector fields are in $\mathbb{P}$. Further, for paths $\mathfrak{p}_{i_1,i_r}$ with $\ell(\mathfrak{p}_{i_1,i_r}) \in \{2,3,4,6\}$ it also follows from the definition of the projection that $\tilde{R}_\psi(\mathfrak{p}_{i_1,i_r}) \in \mathbb{P}$. Suppose now that the result holds true for all paths $\mathfrak{p}$ with $\ell(\mathfrak{p}) = \bar{\ell}$, where $\bar{\ell} \geq 2$, and consider a path $\mathfrak{p}_{i_1,i_r}$ with $\ell(\mathfrak{p}_{i_1,i_r}) = \bar{\ell} + 1$. Observe that all subbrackets of $\tilde{R}_\psi(\mathfrak{p}_{i_1,i_r})$ are in $\mathbb{P}$ by the induction hypothesis and hence, by [PH3], [PH4.a], [PH4.b], we have $\tilde{R}_\psi(\mathfrak{p}_{i_1,i_r}) \in \mathbb{P}$ if

$$
\delta(\mathrm{left}(\tilde{R}_\psi(\mathfrak{p}_{i_1,i_r}))) < \delta(\mathrm{right}(\tilde{R}_\psi(\mathfrak{p}_{i_1,i_r}))) \tag{B.18}
$$

$$
\delta(\mathrm{left}(\mathrm{right}(\tilde{R}_\psi(\mathfrak{p}_{i_1,i_r})))) < \delta(\mathrm{left}(\tilde{R}_\psi(\mathfrak{p}_{i_1,i_r}))); \tag{B.19}
$$

we will show next that these conditions are fulfilled for the above choice of subpaths. By (2.102) and (2.103a) we have that

$$\delta\big(\text{right}(\tilde{R}_\psi(\mathfrak{p}_{i_1,i_r}))\big) = \delta\big(\tilde{R}_\psi(\mathfrak{q})\big) = \ell(q)$$
$$\delta\big(\text{left}(\tilde{R}_\psi(\mathfrak{p}_{i_1,i_r}))\big) = \delta\big(\tilde{R}_\psi(\mathfrak{q}^c)\big) = \ell(\mathfrak{p}_{i_1,i_r}) - \ell(\mathfrak{q}).$$

Since $\lfloor \frac{a}{b} \rfloor \geq \frac{a-b+1}{b}$, for all $a \in \mathbb{Z}, b \in \mathbb{N}$, we have that

$$\ell(q) = \theta(\mathfrak{p}_{i_1,i_r}) - 1 \geq \tfrac{\ell(\mathfrak{p}_{i_1,i_r})+1}{2}, \tag{B.20}$$

for $\ell(\mathfrak{p}_{i_1,i_r}) \geq 5$, and hence we obtain

$$\begin{aligned}\delta\big(\text{right}(\tilde{R}_\psi(\mathfrak{p}_{i_1,i_r}))\big) - \delta\big(\text{left}(\tilde{R}_\psi(\mathfrak{p}_{i_1,i_r}))\big)\\ \geq \ell(\mathfrak{p}_{i_1,i_r}) + 1 - \ell(\mathfrak{p}_{i_1,i_r}) > 0.\end{aligned} \tag{B.21}$$

Thus, (B.18) holds. For (B.19), we first note that

$$\delta\big(\text{left}(\text{right}(\tilde{R}_\psi(\mathfrak{p}_{i_1,i_r})))\big) = \delta\big(\text{left}(\tilde{R}_\psi(\mathfrak{q}))\big) \tag{B.22}$$

and, since $\text{left}(\tilde{R}_\psi(\mathfrak{q})) \in \mathbb{P}$ by the induction hypothesis, it is

$$\delta\big(\text{left}(\tilde{R}_\psi(\mathfrak{q}))\big) \leq \delta\big(\text{right}(\tilde{R}_\psi(\mathfrak{q}))\big) = \ell(q) - \delta\big(\text{left}(\tilde{R}_\psi(\mathfrak{q}))\big) \tag{B.23}$$

according to [PH4.a]. Hence, we obtain

$$\delta\big(\text{left}(\text{right}(\tilde{R}_\psi(\mathfrak{p}_{i_1,i_r})))\big) \leq \tfrac{\ell(\mathfrak{q})}{2}. \tag{B.24}$$

As a result, (B.19) is fulfilled when

$$\tfrac{\ell(q)}{2} \leq \ell(\mathfrak{p}_{i_1,i_r}) - \ell(\mathfrak{q}). \tag{B.25}$$

We now compute

$$\tfrac{3}{2}\ell(q) = \tfrac{3}{2}\lfloor \tfrac{\ell(\mathfrak{p}_{i_1,i_r})}{2} \rfloor + \tfrac{3}{2} \leq \tfrac{3}{4}\ell(\mathfrak{p}_{i_1,i_r}) + \tfrac{3}{2} \leq \ell(\mathfrak{p}_{i_1,i_r}),$$

for $\ell(\mathfrak{p}_{i_1,i_r}) \geq 6$; for $\ell(\mathfrak{p}_{i_1,i_r}) = 5$, we have that $\frac{3}{2}\ell(\mathfrak{q}) = \frac{9}{2} < \ell(\mathfrak{p}_{i_1,i_r})$, thus (B.25) holds for all considered $\mathfrak{p}_{i_1,i_r}$ which proves that (B.19) holds; this concludes the proof.

B.1.6 Proof of Proposition 2

It is clear that (2.115) holds for $\ell(\mathfrak{p}) = 2$, since $B_j(\mathfrak{p})$ is a bracket in $\mathbb{P}$ of degree two, i.e., a bracket of the form $[\phi_{k_1}, \phi_{k_2}]$ with $k_1 < k_2$, and the bracket $[\phi_{k_2}, \phi_{k_1}]$ is not in $\mathbb{P}$. Consider now a path $\mathfrak{p}_{i_1,i_4} = \langle i_1 \mid i_2 \mid i_3 \mid i_4 \rangle$ of length three. For the sake of a shorter notation, we let

$$\phi_{k_j} = h_{i_j,\psi(\mathfrak{p}_{i_j,i_{j+1}})}, \qquad j = 1,2,3. \tag{B.26}$$

We note that (2.114) implies that $k_2 > k_1$ or $k_2 > k_3$, i.e., $k_2 \neq \min_{i=1,2,3} k_i$. Keeping this in mind and following Remark 5, we then obtain

$$B_j(\mathfrak{p}_{i_1,i_4}) = \mathrm{proj}_{\mathbb{P}}\left([\phi_{k_1}, [\phi_{k_2}, \phi_{k_3}]]\right) = \begin{cases} [\phi_{k_2}, [\phi_{k_1}, \phi_{k_3}]] - [\phi_{k_3}, [\phi_{k_1}, \phi_{k_2}]] & \text{if } k_1 = \min\limits_{i=1,2,3} k_i, \\ -[\phi_{k_1}, [\phi_{k_3}, \phi_{k_2}]] & \text{if } k_3 = \min\limits_{i=1,2,3} k_i. \end{cases} \tag{B.27}$$

Further, utilizing the associativity property discussed after Lemma 1, we obtain

$$[\phi_{k_1}, [\phi_{k_2}, \phi_{k_3}]](x) = [[\phi_{k_1}, \phi_{k_2}], \phi_{k_3}](x) = -[\phi_{k_3}, [\phi_{k_1}, \phi_{k_2}]](x) \tag{B.28}$$

Then, by the Jacobi-identity (A.7), we have

$$[\phi_{k_1}, [\phi_{k_2}, \phi_{k_3}]](x) + [\phi_{k_2}, [\phi_{k_3}, \phi_{k_1}]](x) + [\phi_{k_3}, [\phi_{k_1}, \phi_{k_2}]](x) = 0; \tag{B.29}$$

hence, together with (B.28), we infer that

$$[\phi_{k_2}, [\phi_{k_3}, \phi_{k_1}]](x) = 0. \tag{B.30}$$

Thus,

$$B_j(\mathfrak{p}_{i_1,i_4}) = \begin{cases} -[\phi_{k_3}, [\phi_{k_1}, \phi_{k_2}]] & \text{if } k_1 = \min\limits_{i=1,2,3} k_i, \\ -[\phi_{k_1}, [\phi_{k_3}, \phi_{k_2}]] & \text{if } k_3 = \min\limits_{i=1,2,3} k_i. \end{cases} \tag{B.31}$$

The only equivalent bracket is then given by $B_j(\mathfrak{p}_{i_1,i_4}) \sim [\phi_{k_2}, [\phi_{k_1}, \phi_{k_3}]]$ if $k_1 = \min_{i=1,2,3} k_i$ and $B_j(\mathfrak{p}_{i_1,i_4}) \sim [\phi_{k_2}, [\phi_{k_3}, \phi_{k_1}]]$ if $k_3 = \min_{i=1,2,3} k_i$. However, by (B.30), in both cases the equivalent bracket evaluates to zero, thus concluding the proof.

B.2 Proofs of Chapter 3

B.2.1 Proof of Theorem 3

We first show sufficiency and then necessity.
A) Sufficiency. Let some $D^{\dagger}$ be given that fulfills (3.4) and let $x^{\star} = D^{\dagger}z^{\star}$. Then $x^{\star}$ fulfills $Dx^{\star} = Cx^{\star} = z^{\star}$ by (3.4a), (3.4b). The condition that (3.2) has an equilibrium at $x^{\star} = D^{\dagger}z^{\star}$ then amounts to

$$D^{\dagger}z^{\star} = AD^{\dagger}z^{\star} + B\nabla H(CD^{\dagger}z^{\star}). \tag{B.32}$$

Using (3.4b), (3.4c), this holds if

$$0 = B\nabla H(z^{\star}), \tag{B.33}$$

which is fulfilled since $z^\star$ is the minimizer of H.
B) Necessity. Suppose that (3.2) has an equilbrium at $x^\star$ with the property $Dx^\star = Cx^\star = z^\star$ for any $H \in \mathcal{S}_{m,L}$. This implies that

$$x^\star = Ax^\star + B\nabla H(Cx^\star) = Ax^\star \tag{B.34}$$

since $Cx^\star = z^\star$ is the unique minimizer of H. Hence,

$$z^\star = CA^i x^\star = DA^i x^\star \tag{B.35}$$

for any $i = 0, 1, \dots,$ and we infer that

$$\begin{bmatrix} I \\ I \\ I \\ \vdots \\ I \end{bmatrix} z^\star = \underbrace{\begin{bmatrix} C \\ CA \\ CA^2 \\ \vdots \\ CA^{np-1} \end{bmatrix}}_{=:Q_C} x^\star = \underbrace{\begin{bmatrix} D \\ DA \\ DA^2 \\ \vdots \\ DA^{np-1} \end{bmatrix}}_{=:Q_D} x^\star. \tag{B.36}$$

Since the pair (A, C) or the pair (A, D) is observable by assumption, the matrix $Q_C \in \mathbb{R}^{np^2 \times p}$ or $Q_D \in \mathbb{R}^{np^2 \times p}$ has full rank and we infer that there exists some matrix $U \in \mathbb{R}^{np \times p}$ independent of $x^\star, z^\star$ such that $x^\star = Uz^\star$. We next show that $D^\dagger = U$ fulfills (3.4). To this end, note that $z^\star = Cx^\star = CUz^\star$ and, equally well, $z^\star = Dx^\star = DUz^\star$. Since these two equation need to hold for all $z^\star$, this implies that $CU = DU = I$ and we infer that (3.4a), (3.4b) hold for $D^\dagger = U$. The third condition (3.4c) follows directly from (B.34) using $x^\star = Uz^\star$.

B.2.2 Proof of Theorem 4

We follow the lines of the proof of Veenman et al. (2016, Corollary 3) and first show robust stability. Let Π_p be partioned according to (3.10) and observe that (3.11) implies that

$$\begin{bmatrix} G_{yw} \\ I \end{bmatrix}^* \Pi \begin{bmatrix} G_{yw} \\ I \end{bmatrix} + G_{y_\mathrm{p}w}^* \Pi_{\mathrm{p},22} G_{y_\mathrm{p}w} \overset{\mathrm{T}}{\prec} 0. \tag{B.37}$$

Since $\Pi_{\mathrm{p},22} \succeq 0$ according to (3.10), we have

$$\begin{bmatrix} G_{yw} \\ I \end{bmatrix}^* \Pi \begin{bmatrix} G_{yw} \\ I \end{bmatrix} \overset{\mathrm{T}}{\prec} 0. \tag{B.38}$$

Thus, with Assumption 4 and 1., 2., robust stability follows from the standard IQC-Theorem, see, e.g., Fetzer and Scherer (2017); Kao (2012). For robust performance, let $w_\mathrm{p} \in \ell_2^{n_{w_\mathrm{p}}}$ and let w denote the corresponding signal resulting from the feedback interconnection (3.5).

Observe that $w \in \ell_2^{n_w}$ due to robust stability which implies that $y \in \ell_2^{n_y}$. Hence, the z-transforms of w and w_p exist almost everywhere on the unit circle and multiplying (3.11) from left by $\begin{bmatrix}\widehat{w}(z)^* & \widehat{w}_p(z)^*\end{bmatrix}$ and from right by its transposed we obtain

$$\begin{bmatrix}\star\end{bmatrix}^* \Pi_2 \begin{bmatrix}\widehat{y} \\ \widehat{\Delta(y)}\end{bmatrix} + \begin{bmatrix}\star\end{bmatrix}^* \Pi_p \begin{bmatrix}\widehat{w_p} \\ \widehat{y_p}\end{bmatrix} \overset{\mathbb{T}}{\preceq} 0. \tag{B.39}$$

Integrating on both sides over the set $\mathbb{T}$ yields

$$\mathrm{IQC}(\Pi_2, y, \Delta(y)) + \mathrm{IQC}(\Pi_p, w_p, y_p) \leq 0. \tag{B.40}$$

Since $\mathrm{IQC}_1(\Pi_2, y, \Delta(y)) \geq 0$ by assumption, we conclude that the performance criterion defined by Π_p is fulfilled.

B.2.3 Proof of Theorem 5

We first need to shift the non-zero initial conditions of (3.21) to the signal $\tilde{y}_{in}$ in Figure 3.4. To this end, let $\tilde{y}_k$, $k \in \mathbb{N}$, denote the output of (3.21) for $\tilde{\xi}_0 = \xi_0$. We then have that $\tilde{y}_k$ also is the output of

$$\tilde{x}_{t+1} = \rho^{-1}(A + mBC)\tilde{x}_t + \rho^{-1}B\tilde{w}_t \tag{B.41a}$$

$$\tilde{y}_t = C\tilde{x}_t + \tilde{y}_{in,t} \tag{B.41b}$$

$$\tilde{w}_t = \rho^{-t}\Delta_H(\rho^t\tilde{y}_t) \tag{B.41c}$$

with initial condition $\tilde{x}_0 = 0$ and

$$\tilde{y}_{in,t} = \rho^{-t}C(A + mBC)^t\tilde{\xi}_0, \tag{B.42}$$

$t \in \mathbb{N}$. We infer that $\tilde{y}_{in} \in \ell_2^p$ since $\rho^{-1}(A + mBC)$ has all eigenvalues in the open unit disk. We are now ready to apply Theorem 4. Robust performance follows directly from Theorem 4. For robust exponential stability, we note that, by Theorem 4, (3.29a) implies robust stability of the transformed loop from Figure 3.4 against $\mathbf{\Delta}_\rho(m, L)$. Hence, $\tilde{y}$ and $\tilde{w}$ in (3.21) reside in ℓ_2. Together with the assumption that all eigenvalues of $A + mBC$ are located in the open disk of radius ρ, we infer that also $\tilde{\xi} \in \ell_2^{np}$ in (3.21). With $y = \rho_+(\tilde{y})$, $w = \rho_+(\tilde{w})$, $\xi = \rho_+(\tilde{\xi})$, and since ρ_+ maps ℓ_2-signals to $\ell_{2,\rho}$-signals, we conclude that y, w, ξ in (3.15) reside in $\ell_{2,\rho}$, hence the exponential decay follows.

B.2.4 Proof of Lemma 6

The proof boils down to extending the result for monotone nonlinearities from Mancera and Safonov (2005, Lemma 2) to the case of slope-restricted nonlinearities, see Definition 16 and (A.31) for a definition of such nonlinearities and operators. The following argumentation is similar to the one provided in the proof of (D'Amato, Rotea, Megretski, & Jönsson,

2001, Theorem) where only scalar nonlinearities are considered. If M adheres to the conditions given in Lemma 6, by Mancera and Safonov (2005, Lemma 2) it is then known that for any monotone operator $\varphi : \ell_2^p \to \ell_2^p$ we have

$$\langle O_T y, M^\top \varphi(O_T y)\rangle \geq 0 \tag{B.43}$$

for all $y \in \ell_2^p$, $T \in \mathbb{N}$. To use this result, we next rewrite (3.32) in the form (B.43). To this end, consider (3.32), (3.33) with L replaced by $\tilde{L} := L + \varepsilon$, $\varepsilon > 0$. By simple calculations as well as the fact that Δ_H is static and time-invariant, (3.32) is

$$2\langle (\tilde{L} - m)O_T y - \Delta_H(O_T y), M^\top \Delta_H(O_T y)\rangle \geq 0. \tag{B.44}$$

In view of (B.43), for any $y \in \ell_2^p$ we define the auxiliary sequence $\tilde{y} \in \ell_2^p$ by

$$\tilde{y}_t := (\tilde{L} - m)y_t - \Delta_H(y_t), \quad t \in \mathbb{N}. \tag{B.45}$$

Let $\phi : \mathbb{R}^p \to \mathbb{R}$ be defined as

$$\phi(y) = \tfrac{1}{2}(\tilde{L} - m)y^\top y - \left(H(y) - \tfrac{1}{2}my^\top y\right) = \tfrac{1}{2}\tilde{L}y^\top y - H(y), \tag{B.46}$$

which gives $\tilde{y} = \nabla\phi(y)$ and

$$\Delta_H(y) = \nabla\phi(y) - (\tilde{L} - m)y. \tag{B.47}$$

Then (B.44) reads as

$$2\langle O_T\tilde{y}, M^\top \Delta_H(O_T y)\rangle \geq 0. \tag{B.48}$$

Note that $H \in \mathcal{S}_{m,L}$ implies $\phi \in \mathcal{S}_{\varepsilon,\tilde{L}-m}$; we then infer from Rockafellar (1970, Theorem 26.6, Lemma 26.7) that $\nabla\phi$ has a well-defined inverse $(\nabla\phi)^{-1}$ and $(\nabla\phi)^{-1}$ is itself the gradient of a strictly convex function, namely the Legendre conjugate of ϕ, cf. Rockafellar (1970, Theorem 26.5). Thus, we can write (B.48) as

$$2\langle O_T\tilde{y}, M^\top \Delta_H((\nabla\phi)^{-1}(O_T\tilde{y}))\rangle \geq 0. \tag{B.49}$$

Comparing with (B.43), we aim to show that the operator $\varphi := \Delta_H \circ (\nabla\phi)^{-1}$ is monotone. To this end, first note that, since $(\nabla\phi)^{-1}$ is the gradient of a strictly convex function, there exists some function $F : \mathbb{R}^p \to \mathbb{R}$ such that $\left(\varphi(\tilde{y})\right)_t = \nabla F(\tilde{y}_t)$. We next show that ∇F fulfills (A.28), hence $\tilde{w}$ is a monotone operator. By (B.47) we infer that

$$\varphi(\tilde{y}) = \Delta_H\Big((\nabla\phi)^{-1}(\tilde{y})\Big) = \tilde{y} - (\tilde{L} - m)\left(\nabla\phi\right)^{-1}(\tilde{y}). \tag{B.50}$$

Thus, the Hessian of F is given by

$$\nabla^2 F(\tilde{y}_t) = (\tilde{L} - m)\Big(\nabla^2\phi((\nabla\phi)^{-1}(\tilde{y}_t))\Big)^{-1} - I. \tag{B.51}$$

Now, since $\phi \in \mathcal{S}_{\varepsilon,\tilde{L}-m}$, we have by Nesterov (2004, Theorem 2.1.6, Theorem 2.1.11) that $\varepsilon I \preceq \nabla^2\phi(y) \preceq (L - m + \varepsilon)I$ for all $y \in \mathbb{R}^p$, and hence $\nabla^2 F(\tilde{y}_t) \succeq (\frac{\tilde{L}-m}{\tilde{L}-m} - 1)I = 0$ for all $\tilde{y}_t \in \mathbb{R}^p$. Consequently, F is convex ((Nesterov, 2004, Theorem 2.1.4)) and we conclude that φ is a monotone operator, thus (B.44) holds for all $y \in \ell_2^p$, $T \in \mathbb{N}$, $\varepsilon > 0$. For arbitrary but fixed $y \in \ell_2^p$, $T \in \mathbb{N}$, we can then take the limit $\varepsilon \to 0$ in (B.44) since the left-hand side is continuous in ε and infer that (3.32) holds, thus concluding the proof.

B.2.5 Proof of Lemma 7

Let $\phi_T = O_T \phi O_T$, $T \in \mathbb{N}$. Since $\rho_+(\ell_2^p) = \ell_{2,\rho}^p$, we infer from (3.35) that

$$\left\langle \begin{bmatrix} \rho_+(y) \\ \Delta(\rho_+(y)) \end{bmatrix}, \phi_T \begin{bmatrix} \rho_+(y) \\ \Delta(\rho_+(y)) \end{bmatrix} \right\rangle \geq 0 \tag{B.52}$$

for all $y \in \ell_2^p$, $\Delta \in \mathbf{\Delta}$, $T \in \mathbb{N}$. With $\rho_+ \circ \rho_- = \mathrm{id}_{\ell_{2,\rho}}$ this is equivalently formulated as

$$\begin{aligned} &\left\langle \begin{bmatrix} \rho_+(y) \\ \rho_+\Big(\rho_-(\Delta(\rho_+(y)))\Big) \end{bmatrix}, \phi_T \begin{bmatrix} \rho_+(y) \\ \rho_+\Big(\rho_-(\Delta(\rho_+(y)))\Big) \end{bmatrix} \right\rangle \\ &= \left\langle \begin{bmatrix} y \\ \Delta_\rho(y) \end{bmatrix}, \rho_+\Big(\phi_T(\rho_+(\begin{bmatrix} y \\ \Delta_\rho(y) \end{bmatrix}))\Big) \right\rangle \\ &= \left\langle \begin{bmatrix} y \\ \Delta_\rho(y) \end{bmatrix}, O_T\Big(\tilde{\phi}(O_T(\begin{bmatrix} y \\ \Delta_\rho(y) \end{bmatrix}))\Big) \right\rangle \geq 0 \end{aligned} \tag{B.53}$$

for all $y \in \ell_2^p$, $\Delta \in \mathbf{\Delta}$, $T \in \mathbb{N}$. Since $\Delta(\ell_{2,\rho}^p) \subset \ell_{2,\rho}^p$, we have $\Delta_\rho : \ell_2^p \to \ell_2^p$; hence, by the assumption that $\tilde{\phi}$ is bounded on ℓ_2^p, we can take the limit $T \to \infty$ in (B.53) to obtain that (3.36) holds.

B.2.6 Proof of Lemma 8

Observe that $M_\rho = \bar{M}_\rho \otimes I_p$ with

$$\bar{M}_\rho = \mathrm{Toep}(\begin{bmatrix} m_{-\ell_-} & m_{-\ell_-+1} & \dots & m_{\ell_+} \end{bmatrix}) \tag{B.54}$$

and define $M = \rho_- M_\rho \rho_- = (\rho_- \bar{M}_\rho \rho_-) \otimes I_p$. We compute

$$\rho_- \bar{M}_\rho \rho_- = \begin{bmatrix} \bar{m}_0 & \rho^{-1}\bar{m}_1 & \rho^{-2}\bar{m}_2 & \dots \\ \rho^{-1}\bar{m}_{-1} & \rho^{-2}\bar{m}_0 & \rho^{-3}\bar{m}_1 & \dots \\ \ddots & \ddots & \ddots & \ddots \end{bmatrix}. \tag{B.55}$$

Thus, $\rho_- \bar{M}_\rho \rho_- \in \mathbf{M}$ if, for each finite $k \in \mathbb{N}$,

$$\rho^{-k-i}\bar{m}_i \leq 0 \text{ for } i \in \{-\ell_-, \dots, \ell_+\}, i \neq 0, \tag{B.56a}$$

$$\rho^{-2k} \sum_{i=-\ell_-}^{\ell_+} \bar{m}_i \rho^{-i} \geq 0, \rho^{-2k} \sum_{i=-\ell_-}^{\ell_+} \bar{m}_i \rho^{i} \geq 0. \tag{B.56b}$$

With $\rho > 0$, we can get rid of the dependency on k such that the above conditions are equivalent to requiring that the Toeplitz operator in (3.40) is doubly hyperdominant. We hence conclude by Lemma 6 that under these conditions the inequality (3.35) holds with

$$\phi = W^\top \begin{bmatrix} 0 & M^\top \\ M & 0 \end{bmatrix} W, \tag{B.57}$$

where W is defined in (3.33) and $M = (\rho_- \bar{M}_\rho \rho_-) \otimes I_p$. Additionally, $\rho_+ \circ \phi \circ \rho_+$ is bounded on ℓ_2^{2p}, since M is bounded on ℓ_2^p by assumption. Hence, using also that $\Delta_{H,\rho} : \ell_2^p \to \ell_2^p$ as discussed in Section 3.2.4, we may apply Lemma 7 and conclude that

$$\begin{aligned}
&\left\langle \begin{bmatrix} y \\ \Delta_{H,\rho}(y) \end{bmatrix}, \rho_+ W^\top \begin{bmatrix} 0 & M^\top \\ M & 0 \end{bmatrix} W \rho_+ \begin{bmatrix} y \\ \Delta_{H,\rho}(y) \end{bmatrix} \right\rangle \\
&= \left\langle \begin{bmatrix} y \\ \Delta_{H,\rho}(y) \end{bmatrix}, W^\top \begin{bmatrix} 0 & M_\rho{}^\top \\ M_\rho & 0 \end{bmatrix} W \begin{bmatrix} y \\ \Delta_{H,\rho}(y) \end{bmatrix} \right\rangle \geq 0 \qquad \text{(B.58)}
\end{aligned}$$

for any $y \in \ell_2^p$, $\Delta_{H,\rho} \in \mathbf{\Delta}_\rho(m, L)$, thus concluding the proof.

B.2.7 Proof of Theorem 7

The result virtually is a direct consequence of Theorem 5 employing the class of multipliers introduced in Theorem 6. We first note that under the assumption that $A + mBC$ has all eigenvalues in the open disk of radius ρ, the same holds for A_c by (3.54) since ψ_Δ has all its poles at zero. With $\psi_\Delta^* M_\Delta(M_+, M_-, M_0)\psi_\Delta = \Pi$ and by the KYP-Lemma we then observe that (3.55a) is equivalent to the frequency domain inequality (3.29a) for exponential stability in Theorem 5. Further, by Theorem 6 and (3.55b), we infer that $\mathrm{IQC}(\Pi_\rho, \tilde{y}, \Delta_{H,\rho}(\tilde{y})) \geq 0$ holds for all $\Pi_\rho \in \mathbf{\Pi}_{\Delta,\rho}^p$ and all $\Delta_{H,\rho} \in \mathbf{\Delta}_\rho(m, L)$. Hence, by Theorem 5, we conclude that the origin is globally robustly exponentially stable against $\mathbf{\Delta}(m, L)$ with rate ρ for (3.15), which in turn implies the last claim in Theorem 7, thus concluding the proof.

B.2.8 Proof of Dimensionality Reduction in Theorem 7

We follow the arguments of Lessard et al. (2016) to prove this statement. It is clear that if (3.56a) has a solution $\bar{P}, m_+, , m_-, m_0$, then $P = \overline{P} \otimes I_p, M_+ = m_+ \otimes I_p, m_- \otimes I_p, m_0 \otimes I_p$. Now suppose that (3.55a) has a solution P, M_+, M_-, M_0. Multiply (3.55a) from right and left by $\mathrm{blkdiag}(I_{n_c} \otimes e_1, I_{n_c} \otimes e_1)$ and its transpose, where $e_1 \in \mathbb{R}^{p \times 1}$ is the first unit vector. Observing that for any $U = \bar{U} \otimes I_p, \bar{U} \in \mathbb{R}^{n_c \times r}$, we have $U(I_n \otimes e_1) = (\bar{U} \otimes I_p)(I_{n_c} \otimes e_1) = \bar{U} \otimes e_1 = (I_r \otimes e_1)(\bar{U} \otimes 1)$, we obtain

$$\begin{bmatrix} A_c & B_c \\ I & 0 \end{bmatrix}^\top \begin{bmatrix} P & 0 \\ 0 & -\bar{P} \end{bmatrix} \begin{bmatrix} A_c & B_c \\ I & 0 \end{bmatrix} + \begin{bmatrix} \bar{C}_c & \bar{D}_c \end{bmatrix}^\top (I_{3p} \otimes e_1)^\top M_\Delta (I_{3p} \otimes e_1) \begin{bmatrix} \bar{C}_c & \bar{D}_c \end{bmatrix} \prec 0 \quad \text{(B.59)}$$

with $\bar{P} = (I_n \otimes e_1)^\top P (I_n \otimes e_1)$. Let $M_\Delta(M_+, M_-, M_0)$ be defined as in (3.52) explicitly including the dependency on M_+, M_-, M_0, and note that

$$\begin{aligned}
&(I_{3p} \otimes e_1)^\top M_\Delta(M_+, M_-, M_0)(I_{3p} \otimes e_1) \\
&= (I_{3p} \otimes e_1)^\top (M_\Delta(m_+, m_-, m_0) \otimes I_p)(I_{3p} \otimes e_1) \\
&= (I_{3p} \otimes e_1)^\top (M_\Delta(m_+, m_-, m_0) \otimes e_1) \\
&= M_\Delta(m_+, m_-, m_0). \qquad \text{(B.60)}
\end{aligned}$$

Hence, $\bar{P} = (I_n \otimes e_1)^\top P (I_n \otimes e_1), m_+, m_-, m_0$ is a solution to (3.56a) which concludes the proof.

B.2.9 Proof of Theorem 8

We first observe that robust stability follows directly from Theorem 7 noting that $\boldsymbol{A}_c = A_c(1)$, $\boldsymbol{B}_{c,1} = B_c$, $\boldsymbol{C}_{c,1} = C_c(1)$, $\boldsymbol{D}_{c,11} = D_c$ by (C.29) and $\boldsymbol{C}_{c,1}^\top \boldsymbol{C}_{c,1} \succeq 0$; hence (3.64a) implies that (3.55) holds with $\rho = 1$. Consider the complete transfer matrix ψ_c as defined in (3.61) and let a state-space realization be given as in (3.62) specifically stated in (C.29) in Appendix C.2.2. Let x_c denote the solution of

$$x_{c,t+1} = \boldsymbol{A}_c x_{c,t} + \boldsymbol{B}_{c,1} w_t + \boldsymbol{B}_{c,2} w_{p,t} \tag{B.61a}$$

$$w_t = \Delta(CN^\top x_{c,t}) \tag{B.61b}$$

with initial condition $x_{c,0} = 0$. Note that with $\bar{x}_{c,t} = N^\top x_{c,t}$ we have

$$\bar{x}_{c,t+1} = (A + mBC)\bar{x}_{c,t} + B\Delta(C\bar{x}_{c,t}) + B_p w_{p,t}, \tag{B.62}$$

i.e., $\bar{x}_{c,t}$ follows the same dynamics as (3.15) and the performance output is given by $y_{p,t} = C_p \bar{x}_{c,t} = \boldsymbol{C}_{c,2} x_{c,t}$. Consequently, multiplying (3.64a) from left by $\begin{bmatrix} x_{c,t}^\top & w_t^\top \end{bmatrix}$ and from right by its transpose, we infer

$$(\star)^\top P_p(\boldsymbol{A}_c x_{c,t} + \boldsymbol{B}_{c,1} w_t) - x_{c,t}^\top P_p x_{c,t} \leq -y_{p,t}^\top y_{p,t} - (\star)^\top M_\Delta(\boldsymbol{C}_{c,1} x_{c,t} + \boldsymbol{D}_{c,11} w_t). \tag{B.63}$$

Let

$$P_p' = N^\top P_p N, \qquad P_p'' = P_p - NN^\top P_p NN^\top, \tag{B.64}$$

i.e., $P_p = P_p'' + NP_p'N^\top$ and $N^\top P_p'' N = 0$. By (C.29), we have $\boldsymbol{B}_{c,2} = NB_p$; hence we infer that $\boldsymbol{B}_{c,2}{}^\top P_p \boldsymbol{B}_{c,2} = \boldsymbol{B}_{c,2}^\top N P_p' N^\top \boldsymbol{B}_{c,2}$. We then calculate

$$\begin{aligned} & x_{c,t+1}^\top P_p x_{c,t+1} - x_{c,t}^\top P_p x_{c,t} \\ &= (\star)^\top P_p(\boldsymbol{A}_c x_{c,t} + \boldsymbol{B}_{c,1} w_t) - x_{c,t}^\top P_p x_{c,t} + 2x_{c,t+1}^\top P_p \boldsymbol{B}_{c,2} w_{p,t} \\ &\quad + w_{p,t}^\top \boldsymbol{B}_{c,2}^\top N P_p' N^\top \boldsymbol{B}_{c,2} w_{p,t}. \end{aligned} \tag{B.65}$$

Using (B.63), (B.65) and summing from $t = 0$ to $t_{\max} \in \mathbb{N}$, we obtain

$$\begin{aligned} \sum_{t=0}^{t_{\max}} x_{c,t+1}^\top P_p x_{c,t+1} - x_{c,t}^\top P_p x_{c,t} \leq \sum_{t=0}^{t_{\max}} \Big(& -y_{p,t}^\top y_{p,t} - (\star)^\top M_\Delta(\boldsymbol{C}_{c,1} x_{c,t} + \boldsymbol{D}_{c,11} w_t) \\ & + 2x_{c,t+1}^\top P_p \boldsymbol{B}_{c,2} w_{p,t} \\ & + w_{p,t}^\top \boldsymbol{B}_{c,2}^\top N P_p' N^\top \boldsymbol{B}_{c,2} w_{p,t} \Big). \end{aligned} \tag{B.66}$$

We note that $w_{p,t}$ is a discrete-time white noise process with independent components; hence, $w_{p,t}$ and $x_{c,t+1}$ are independent for any $t \in \mathbb{N}$, $\mathrm{E}(w_{p,t}) = 0$ and $\mathrm{E}(w_{p,t}^\top X w_{p,t}) = \mathrm{tr}(X)$ for any $X \in \mathbb{R}^{n_{w_p} \times n_{w_p}}$. Taking expectations, we hence infer

$$\begin{aligned} \mathrm{E}\Big(\sum_{t=0}^{t_{\max}} x_{c,t+1}^\top P_p x_{c,t+1} - x_{c,t}^\top P_p x_{c,t}\Big) \leq \mathrm{E}\Big(\sum_{t=0}^{t_{\max}} & -y_{p,t}^\top y_{p,t} - (\star)^\top M_\Delta(\boldsymbol{C}_{c,1} x_{c,t} + \boldsymbol{D}_{c,11} w_t)\Big) \\ & + t_{\max} \mathrm{tr}(\boldsymbol{B}_{c,2}^\top N P_p' N^\top \boldsymbol{B}_{c,2}). \end{aligned} \tag{B.67}$$

With $x_{c,0} = 0$ we further note that for any $t_{\max} \in \mathbb{N}$ we have

$$\sum_{t=0}^{t_{\max}} x_{c,t+1}^{\top} P_p x_{c,t+1} - x_{c,t}^{\top} P_p x_{c,t} = x_{c,t_{\max}+1}^{\top} P_p x_{c,t_{\max}+1}. \tag{B.68}$$

Combining (B.67) with (B.68) and using (3.64b) we then obtain

$$\begin{aligned} &\tfrac{1}{t_{\max}} \sum_{t=0}^{t_{\max}} \mathrm{E}(y_{p,t}^{\top} y_{p,t}) + \tfrac{1}{t_{\max}} \mathrm{E}(x_{c,t_{\max}+1}^{\top} P_p x_{c,t_{\max}+1}) \\ &\leq - \tfrac{1}{t_{\max}} \mathrm{E}\Big(\sum_{t=0}^{t_{\max}} (\star)^{\top} M_{\Delta} (\boldsymbol{C}_{c,1} x_{c,t} + \boldsymbol{D}_{c,11} w_t)\Big) + \gamma^2 \end{aligned} \tag{B.69}$$

Now note that, for a fixed realization $w_{p,t}$, $\sum_{t=0}^{\infty} (\star)^{\top} M_{\Delta} (\boldsymbol{C}_{c,1} x_{c,t} + \boldsymbol{D}_{c,11} w_t)$ is a time-domain representation of $\mathrm{IQC}(\psi_{\Delta} M_{\Delta} \psi_{\Delta}, y, \Delta_H(y))$. Since (3.32) is a hard IQC, we infer that

$$\sum_{t=0}^{t_{\max}} (\star)^{\top} M_{\Delta} (\boldsymbol{C}_{c,1} x_{c,t} + \boldsymbol{D}_{c,11} w_t) \geq 0 \tag{B.70}$$

for any realization $w_{p,t}$ and any $t_{\max}$, and the latter inequality persists to hold after taking expectations. Additionally, with P_p being positive definite, $\mathrm{E}(x_{c,t}^{\top} P_p x_{c,t})$ is positive as well. Thus, if we let $t_{\max}$ tend to infinity, we finally obtain from (B.69)

$$\limsup_{t_{\max} \to \infty} \tfrac{1}{t_{\max}} \sum_{t=0}^{t_{\max}} \mathrm{E}(y_{p,t}^{\top} y_{p,t}) \leq \gamma^2, \tag{B.71}$$

hence concluding the proof.

B.2.10 Proof of Theorem 10

Robust exponential stability follows directly from Theorem 9. For robust performance, we first note that (3.78a), (3.78b) implies that (3.74) holds with $P_{p,22} = P_{22}$, $Q_A = P_{22} A$, $Q_B = P_{22} B$, and making use of the specific state-space realization from Appendix C.2.2, thus (3.78a), (3.78b) implies (3.64a). Note further that U_p as defined in (3.75) is independent of A, B by the specific structure of P_p. We further note that (3.78c) holds if and only if $P_{22} \succ 0$ and $Z \succ Q_B^{\top} P_{22}^{-1} Q_B = B^{\top} P_{22} B$. Together with that the trace condition (3.78d) then implies that $\mathrm{tr}(B^{\top} P_{22} B) < \gamma^2$. Now note that with the specific state-space realization Appendix C.2.2 and by the definition of P_p we have that $\mathrm{tr}(\boldsymbol{B}_{c,2}^{\top} P_p \boldsymbol{B}_{c,2}) = \mathrm{tr}(B^{\top} P_{22} B)$, hence (3.78c) together with (3.78d) implies (3.64b), thus concluding the proof.

B.2.11 Proof of Lemma 11

We prove this result utilizing Lemma 14. By (A.26), we infer that

$$
\begin{aligned}
&\|\nabla H(y_1+y_2)-\nabla H(y_1)-\beta y_2\|^2 \\
&=\|\nabla H(y_1+y_2)-\nabla H(y_1)\|^2-2\beta(\nabla H(y_1+y_2)-\nabla H(y_1))^\top y_2+\beta^2\|y_2\|^2 \\
&\leq \|\nabla H(y_1+y_2)-\nabla H(y_1)\|^2-2\beta(\tfrac{mL}{m+L}\|y_2\|^2+\tfrac{1}{m+L}\|\nabla H(y_1+y_2)-\nabla H(y_2)\|^2)+\beta^2\|y_2\|^2 \\
&=(1-\tfrac{2\beta}{m+L})\|\nabla H(y_1+y_2)-\nabla H(y_1)\|^2+(\beta^2-2\beta\tfrac{mL}{m+L})\|y_2\|^2. \qquad \text{(B.72)}
\end{aligned}
$$

Since $\beta \leq \frac{L+m}{2}$ it is $1-\frac{2\beta}{m+L}\geq 0$ and using the Lipschitz property (A.25b) of the gradient we obtain further

$$
\begin{aligned}
&(1-\tfrac{2\beta}{m+L})\|\nabla H(y_1+y_2)-\nabla H(y_1)\|^2+(\beta^2-2\beta\tfrac{mL}{m+L})\|y_2\|^2 \\
&\leq (1-\tfrac{2\beta}{m+L})L^2\|y_2\|^2+(\beta^2-2\beta\tfrac{mL}{m+L})\|y_2\|^2 \\
&=\frac{(m+L)L^2-2\beta L^2+\beta^2(m+L)-2\beta mL}{m+L}\|y_2\|^2 \\
&=\frac{(m+L)L^2-2\beta L(m+L)+\beta^2(m+L)}{m+L}\|y_2\|^2 \\
&=(L-\beta)^2\|y_2\|^2, \qquad \text{(B.73)}
\end{aligned}
$$

hence

$$
\|\nabla H(y_1+y_2)-\nabla H(y_1)-\beta y_2\|^2 \leq (L-\beta)^2\|y_2\|^2. \qquad \text{(B.74)}
$$

With $y_1=\bar{z}$, $y_2=z$, the claim then follows.

B.2.12 Proof of Lemma 12

We prove this result by showing that the equilibrium $\xi^\star=0$ is globally exponentially stable with rate ρ for the transformed dynamics (3.124) for the considered class of objective functions H. To this end, we write (3.124) as a standard robust state-feedback problem and then use quadratic Lyapunov functions together with an S-procedure argument. We first calculate for the uncertain terms in (3.124)

$$
\begin{aligned}
&\nabla H(C(A_1+I)x_t+CB_1K\xi_t)-\nabla H(Cx_t) \\
&=H_1(C(A_1+I)x_t+CB_1K\xi_t)-H_1Cx_t \\
&+T^\top\nabla H_2(TC(A_1+I)x_t+TCB_1K\xi_t)-T^\top\nabla H_2(TCx_t) \\
&=H_1(CA_1\xi_t+CB_1K\xi_t) \qquad \text{(B.75)}\\
&+T^\top\Big(\nabla H_2(TC(A_1+I)x_t+TCB_1K\xi_t)-\nabla H_2(TCx_t)\Big)
\end{aligned}
$$

and then define the uncertainty as

$$w_t = \nabla H_2(TC(A_1 + I)x_t + TCB_1K\xi_t) - \nabla H_2(TCx_t) - \beta TC(A_1 + B_1K)\xi_t \tag{B.76}$$

with $\beta = \frac{1}{2}(L_2 + m_2)$. By Lemma 11, we then have that

$$\begin{bmatrix} \xi_t \\ w_t \end{bmatrix}^\top \begin{bmatrix} (L_2 - \beta)\Gamma^\top\Gamma & 0 \\ 0 & -\frac{1}{L_2-\beta}I \end{bmatrix} \begin{bmatrix} \xi_t \\ w_t \end{bmatrix} \geq 0, \tag{B.77}$$

for any $\xi_t \in \mathbb{R}^{np}$, where $\Gamma = TC(A_1 + B_1K)$. The transformed dynamics (3.124) then read as

$$\begin{aligned} \xi_{t+1} = &(A_2 + B_2H_1CA_1 + \beta B_2T^\top TCA_1)\xi_t + (B_1 + B_2H_1CB_1 + \beta B_2T^\top TCB_1)K\xi_t \\ &+ B_2T^\top w_t. \end{aligned} \tag{B.78}$$

Using the shorthand notation (3.128), (B.78) can equivalently be written as

$$\xi_{t+1} = (\bar{A} + \bar{B}K)\xi_t + \bar{G}w_t, \tag{B.79}$$

hence the problem of finding parameters K_i is a standard robust state-feedback problem and the proof is standard from now on. Consider a quadratic Lyapunov function candidate $V : \mathbb{R}^{np} \to \mathbb{R}$ defined as $V(\xi) = \xi^\top P\xi$ with $P \succ 0$. For exponential convergence with rate ρ of (B.79) for all w that fulfill (B.77) – and hence exponential convergence with rate ρ of the original algorithm to the minimizer of H for all $H \in \mathcal{S}_{m,L}$ – it is sufficient that there exists a $P \succ 0$ such that

$$V(\xi_{t+1}) - \rho^2 V(\xi_t) = \begin{bmatrix} \star \end{bmatrix}^\top \begin{bmatrix} \star \end{bmatrix}^\top \begin{bmatrix} P & 0 \\ 0 & -\rho^2 P \end{bmatrix} \begin{bmatrix} \bar{A} + \bar{B}K & G \\ I & 0 \end{bmatrix} \begin{bmatrix} \xi_t \\ w_t \end{bmatrix} < 0 \tag{B.80}$$

for all $\begin{bmatrix} \xi_t^\top & w_t^\top \end{bmatrix}^\top \neq 0$ that fulfill (B.77). By the S-procedure (cf. Lemma 18), it is hence sufficient that there exist $P \succ 0$ and $\lambda \geq 0$ such that

$$\begin{bmatrix} \star \end{bmatrix}^\top \begin{bmatrix} P & 0 \\ 0 & -\rho^2 P \end{bmatrix} \begin{bmatrix} \bar{A} + \bar{B}K & \bar{G} \\ I & 0 \end{bmatrix} + \lambda \begin{bmatrix} (L_2 - \beta)\Gamma^\top\Gamma & 0 \\ 0 & -\frac{1}{L_2-\beta}I \end{bmatrix} \prec 0. \tag{B.81}$$

We exclude the case $\lambda = 0$ and thus set $\lambda = 1$ without loss of generality. We can write the above inequality equivalently as

$$\begin{bmatrix} \star \end{bmatrix}^\top \begin{bmatrix} P^{-1} & 0 \\ 0 & I \end{bmatrix} \begin{bmatrix} P(\bar{A} + \bar{B}K) & P\bar{G} \\ \sqrt{L_2 - \beta}\Gamma & 0 \end{bmatrix} - \begin{bmatrix} \rho^2 P & 0 \\ 0 & \frac{1}{L_2-\beta}I \end{bmatrix} \prec 0. \tag{B.82}$$

Since P is assumed to be a positive definite matrix, we can employ Schur complements such that the above matrix inequality holds if and only if

$$\begin{bmatrix} P & 0 & P(\bar{A} + \bar{B}K) & P\bar{G} \\ \star & I & \sqrt{L_2 - \beta}\Gamma & 0 \\ \star & \star & \rho^2 P & 0 \\ \star & \star & \star & \frac{1}{L_2-\beta}I \end{bmatrix} \succ 0. \tag{B.83}$$

We multiply from left and right by the positive definite block diagonal matrix

$$\text{blkdiag}(\sqrt{L_2 - \beta}P^{-1}, I, \sqrt{L_2 - \beta}P^{-1}, \sqrt{L_2 - \beta}I) \tag{B.84}$$

such that we obtain

$$\begin{bmatrix} (L_2 - \beta)P^{-1} & 0 & (L_2 - \beta)(\bar{A} + \bar{B}K)P^{-1} & (L_2 - \beta)\bar{G} \\ \star & I & (L_2 - \beta)\Gamma P^{-1} & 0 \\ \star & \star & \rho^2(L_2 - \beta)P^{-1} & 0 \\ \star & \star & \star & I \end{bmatrix} \succ 0. \tag{B.85}$$

We note that the additional scaling is not required and included for numerical purpose only. The variable transformation $Q = (L_2 - \beta)P^{-1}$, $M = (L_2 - \beta)KP^{-1}$ then gives (3.127), thus concluding the proof.

C
Additional Material

C.1 Additional Material to Chapter 2

C.1.1 A Simplified Procedure for the Construction of Approximating Input Sequences

Our objective in this section is to provide a modified version of the construction procedure from Liu (1997a) using the structural properties of the problem at hand, which leads to considerable simplifications. Given the scopes of this work and the complicated nature of the subject, we do not discuss this procedure in detail; we refer the reader to Michalowsky et al. (2017b), as well as the original work Liu (1997a).

The computation procedure we present in the following applies to extended systems of the form

$$\dot{x} = f_0(x) + \sum_{\substack{B \in \mathcal{B} \\ \delta(B) \geq 2}} v_B B(x), \tag{C.1}$$

where, $f_0 : \mathbb{R}^N \to \mathbb{R}^N$, $f_0 \in \mathcal{C}^\infty$, $v_B \in \mathbb{R}$, and $\mathcal{B} \subset \mathbb{P}$ is a finite set of Lie brackets from a given P. Hall basis $\mathcal{PH}(\Phi) = (\mathbb{P}, \prec)$ with $\Phi = \{\phi_1, \phi_2, \ldots, \phi_M\}$, $\phi_k : \mathbb{R}^N \to \mathbb{R}^N$, $\phi_k \in \mathcal{C}^\infty$. The procedure explicitly exploits the following assumption.

Assumption 5. For any $B \in \mathcal{B}$, $\delta_k(B) \in \{0, 1\}$, $k = 1, 2, \ldots, M$. •

We emphasize that, by construction, this is fulfilled for admissible Lie bracket representations obtained from the procedure presented in Section 2.3. The following procedure computes approximating input sequences U_k^σ such that the solutions of

$$\dot{X}^\sigma = f_0(X^\sigma) + \sum_{k=1}^{M} \phi_k(X^\sigma) U_k^\sigma(t). \tag{C.2}$$

converge to those of (C.1) with increasing σ. An exemplary MATLAB implementation of the following computation procedure is available from Michalowsky et al. (2017b).

Before we present the procedure, we first provide a formal definition of the already mentioned equivalence relation on the set of Lie brackets:

Definition 17 (Equivalent Brackets). Let $\mathcal{PH}(\Phi) = (\mathbb{P}, \prec)$ be a P. Hall basis of the set of vector fields $\Phi = \{\phi_1, \ldots, \phi_M\}$ and let $\delta_k(B)$ denote the degree of the vector field ϕ_k in the bracket $B \in \mathcal{PH}$. We say that two brackets $B_1, B_2 \in \mathbb{P}$ are *equivalent*, denoted by $B_1 \sim B_2$, if $\delta_k(B_1) = \delta_k(B_2)$ for all $k = 1, \ldots, M$. •

For a given set of brackets $\mathbb{P}$, we then denote by $E_B = \{\tilde{B} \in \mathbb{P} : \tilde{B} \sim B\}$ the equivalence class corresponding to the bracket $B \in \mathbb{P}$. Note that, by definition of the equivalence relation, all brackets contained in an equivalence class $E = \{B_1, B_2, \ldots, B_r\}$, $r \in \mathbb{N}_{>0}$, have the same degree, and we hence let $\delta(E) = \delta(B_k)$, $k \in \{1, 2, \ldots, r\}$, denote the degree of the equivalence class. For the construction of the sets of frequencies, we also need the following two definitions:

Definition 18 (Minimally Canceling). A set $\Omega = \{\omega_1, \ldots, \omega_m\}$, $\omega_i \in \mathbb{R}$, is called *minimally canceling* if for each collection of integers $\{y_i\}_{i=1}^m$ such that $\sum_{k=1}^m |y_k| \leq m$ we have $\sum_{k=1}^m y_k \omega_k = 0$ if and only if all y_k are equal. •

Definition 19 (Independent Collection). A finite collection of sets $\{\Omega_\lambda\}_{\lambda=1}^N$, $N \in \mathbb{N}_{>0}$, with $\Omega_\lambda = \{\omega_{\lambda,1}, \omega_{\lambda,2} \ldots, \omega_{\lambda,M_\lambda}\}$, $\omega_{\lambda,i} \in \mathbb{R}$, is called *independent* if the following holds:

1. the sets Ω_λ are pairwise disjoint, and
2. for each collection of integers $\{y_{i,k}\}_{i=1}^N$, $k = 1, \ldots, M_i$, such that

$$\sum_{i=1}^{N} \sum_{k=1}^{M_i} y_{i,k} \omega_{i,k} = 0 \quad \text{and} \quad \sum_{i=1}^{N} \sum_{k=1}^{M_i} |y_{i,k}| \leq \sum_{i=1}^{N} M_i$$

   we have $\sum_{k=1}^{M_i} y_{i,k} \omega_{i,k} = 0$, for each $i = 1, 2, \ldots, N$. •

We next present the computation procedure.

Computation Procedure

Step 1 (Determining the equivalence classes): For all $B \in \mathcal{B}$, determine the associated (reduced) equivalence class

$$E_B = \{\tilde{B} \in \mathbb{P} : \tilde{B} \sim B, \tilde{B}(z) \not\equiv 0\} = \{\tilde{B}_{E,1}, \tilde{B}_{E,2}, \ldots, \tilde{B}_{E,|E(B)|}\},$$

and let $\mathcal{E} = \{E_B, B \in \mathcal{B}\}$. For each $B \in \mathbb{P}$, set

$$\tilde{v}_B = \begin{cases} v_B & \text{if } B \in \mathcal{B} \\ 0 & \text{otherwise.} \end{cases}$$

Step 2 (Determining the frequencies): For all $E \in \mathcal{E}_2 := \{E \in \mathcal{E} : \delta(E) = 2\}$, choose $|\mathcal{E}_2|$ distinct frequencies $\omega_E \in \mathbb{R} \setminus \{0\}$, and for all $E \in \mathcal{E}, \delta(E) \geq 3$ choose $M|E|$ sets

$$\Omega^+_{E,\rho,k} = \begin{cases} \{\omega_{E,\rho,k}\} & \text{if } \delta_k(E) = 1 \\ \varnothing & \text{if } \delta_k(E) = 0 \end{cases}$$
$$\Omega^-_{E,\rho,k} = -\Omega^+_{E,\rho,k},$$

$\omega_{E,\rho,k} \in \mathbb{R} \setminus \{0\}, k = 1, \ldots, M, \rho = 1, \ldots, |E|$, such that

1. For each $E \in \mathcal{E}, \delta(E) \geq 3$, and each $\rho = 1, \ldots, |E|$, the set $\Omega^+_{E,\rho} = \bigcup_{k=1}^M \Omega^+_{E,\rho,k}$ is minimally canceling.
2. The collection of sets

$$\left\{\{\omega_E, -\omega_E\}_{E \in \mathcal{E}_2}, \{\Omega^+_{E,\rho} \cup \Omega^-_{E,\rho}\}_{\substack{E \in \mathcal{E}, \delta(E) \geq 3, \\ \rho = 1, \ldots, |E|}}\right\}$$

   is independent.

Step 3 (Calculating the auxiliary matrix Ξ_E): For all $E \in \mathcal{E}$ with $\delta(E) \geq 3$, compute

$$\Xi_E = \begin{bmatrix} \xi^+_{\tilde{B}_{E,1},1} & \xi^+_{\tilde{B}_{E,1},2} & \cdots & \xi^+_{\tilde{B}_{E,1},|E|} \\ \xi^+_{\tilde{B}_{E,2},1} & \xi^+_{\tilde{B}_{E,2},2} & \cdots & \xi^+_{\tilde{B}_{E,2},|E|} \\ \vdots & \vdots & \ddots & \vdots \\ \xi^+_{\tilde{B}_{E,|E|},1} & \xi^+_{\tilde{B}_{E,|E|},2} & \cdots & \xi^+_{\tilde{B}_{E,|E|},|E|} \end{bmatrix},$$

where, for any $B \in E$, we let

$$\xi^+_{B,\rho} = \hat{g}_B(\omega_{E,\rho,\theta_B(1)}, \omega_{E,\rho,\theta_B(2)}, \ldots, \omega_{E,\rho,\theta_B(\delta(B))}),$$

with $\theta_B(i) = k$ if the ith vector field in B is ϕ_k and where $\hat{g}_B : \mathbb{R}^{\delta(B)} \to \mathbb{R}$ is defined as follows:

- If $\delta(B) = 1$, then $\hat{g}_B(\tilde{\omega}_1) = 1$.
- If $B = [B_1, B_2]$, then

$$\hat{g}_B(\tilde{\omega}_1, \tilde{\omega}_2, \ldots, \tilde{\omega}_{\delta(B)}) = \frac{\hat{g}_{B_1}(\tilde{\omega}_1, \tilde{\omega}_2, \ldots, \tilde{\omega}_{\delta(B_1)})}{\sum_{i=1}^{\delta(B_1)} \tilde{\omega}_i} \hat{g}_{B_2}(\tilde{\omega}_{\delta(B_1)+1}, \tilde{\omega}_{\delta(B_1)+2}, \ldots, \tilde{\omega}_{\delta(B_1)+\delta(B_2)}).$$

Step 4 (Calculating the input coefficients): For all $E \in \mathcal{E}$ with $\delta(E) = 2$, i.e., $E(B) = \{B\} = [\phi_{k_1}, \phi_{k_2}]$, set

$$\eta_{E,k_1}(\omega_E) = \mathrm{i}\frac{1}{\beta_E}\operatorname{sign}(\tilde{v}_B \omega_E)\sqrt{\tfrac{1}{2}|\tilde{v}_B \omega_E|}$$
$$\eta_{E,k_2}(\omega_E) = \beta_E\sqrt{\tfrac{1}{2}|\tilde{v}_B \omega_E|},$$

where $\beta_E \neq 0$. For all $E \in \mathcal{E}$ with $\delta(E) \geq 3$ let[1]

$$\begin{bmatrix} \gamma_{E,1} \\ \gamma_{E,2} \\ \vdots \\ \gamma_{E,|E|} \end{bmatrix} = \Xi_E^{-1} \begin{bmatrix} \tilde{v}_{\tilde{B}_{E,1}} \\ \tilde{v}_{\tilde{B}_{E,2}} \\ \vdots \\ \tilde{v}_{\tilde{B}_{E,|E|}} \end{bmatrix}$$

and compute $\eta_E(\omega)$ as follows:

- If $\delta(E)$ is odd, for each $\rho = 1, \dots, |E|$, take

$$\eta_E(\omega) = \beta_{E,\omega}\left(\tfrac{1}{2}\gamma_{E,\rho}\mathrm{i}^{\delta(E)-1}\right)^{\frac{1}{\delta(E)}}$$

  for all $\omega \in \Omega^+_{E,\rho}$, and

- if $\delta(E)$ is even, for each $\rho = 1, \dots, |E|$, take

$$\eta_E(\tilde{\omega}) = \mathrm{i}\beta_{E,\tilde{\omega}}\mathrm{sign}\left(\gamma_{E,\rho}(t)\mathrm{i}^{\delta(E)-2}\right)\left|\tfrac{1}{2}\gamma_{E,\rho}(t)\mathrm{i}^{\delta(E)-2}\right|^{\frac{1}{\delta(E)}}$$

  for some $\tilde{\omega} \in \Omega^+_{E,\rho}$ and

$$\eta_E(\omega) = \beta_{E,\omega}\left|\tfrac{1}{2}\gamma_{E,\rho}(t)\mathrm{i}^{\delta(E)-2}\right|^{\frac{1}{\delta(E)}}$$

  for all $\omega \in \Omega^+_{E,\rho} \setminus \{\tilde{\omega}\}$.

In both cases $\beta_{E,\omega} \in \mathbb{R}$ can be chosen freely such that they fulfill $\prod_{\omega \in \Omega^+_{E,\rho}} \beta_{E,\omega} = 1$.

Step 5 (Calculating the approximating inputs): Compute the input according to $U_k^\sigma(t) = \sum_{E \in \mathcal{E}} U_{k,E}^\sigma(t)$ with $U_{k,E}^\sigma : \mathbb{R} \to \mathbb{R}$ being defined as

$$U_{k,E}^\sigma(t) = \begin{cases} 0 & \text{if } \delta_k(E) = 0 \\ 2\sqrt{\sigma}\mathrm{Re}\left(\eta_{E,k}(\omega_E)e^{\mathrm{i}\sigma\omega_E t}\right) & \text{if } \delta(E) = 2, \delta_k(E) = 1 \\ 2\sigma^{\frac{N-1}{N}} \sum\limits_{\rho=1}^{|E|} \mathrm{Re}\left(\eta_E(\omega_{E,\rho,k})e^{\mathrm{i}\sigma\omega t}\right) & \text{if } \delta(E) = N, \delta_k(E) = 1. \end{cases}$$

Note that this computation procedure is a reformulation of the one presented in Liu (1997a) (see the supplementary material provided in Michalowsky et al. (2017b) for a derivation) exploiting two structural properties of the problem at hand: (1) each $B \in \mathcal{B}$

[1] We tacitly assume here that Ξ_E is invertible. It has been shown in Liu (1997a) that there always exists a choice of frequencies such that the corresponding matrix obtained when using "full" equivalence classes is invertible; however, it is not clear whether this also holds in the case of reduced equivalence classes where Ξ_E is a submatrix obtained from the general one by removing several rows and columns.

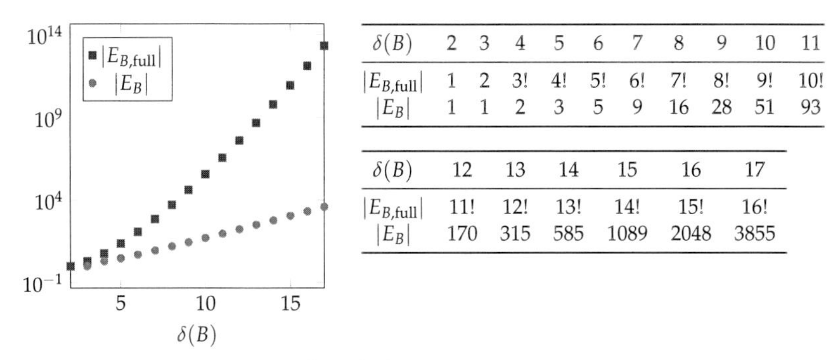

$\delta(B)$	2	3	4	5	6	7	8	9	10	11
$\lvert E_{B,\text{full}}\rvert$	1	2	3!	4!	5!	6!	7!	8!	9!	10!
$\lvert E_B\rvert$	1	1	2	3	5	9	16	28	51	93

$\delta(B)$	12	13	14	15	16	17
$\lvert E_{B,\text{full}}\rvert$	11!	12!	13!	14!	15!	16!
$\lvert E_B\rvert$	170	315	585	1089	2048	3855

Figure C.1. A comparison of $|E_{B,\text{full}}|$ and $|E_B|$ for a specific choice of the P. Hall basis that fulfills the assumptions as in Proposition 2. The numbers were obtained by symbolically computing the resulting vector fields using a computer algebra system. Interestingly, the sequence of $|E_B|$ has two matching sequences *The On-Line Encyclopedia of Integer Sequences, A000048* (n.d.) and *The On-Line Encyclopedia of Integer Sequences, A006788* (n.d.) except for the value for $\delta(B) = 15$ which should be 1091 or 1092, thus we conjecture that these sequences are a good upper bound for $|E_B|$.

fulfills $\delta_k(B) \in \{0,1\}$ for all $k = 1,2,\ldots,M$ (cf. Assumption 5) and (2) a large number of the equivalent brackets evaluate to zero (see Figure C.1). Note that (1) simplifies the calculation of $\xi^{+}_{B,\rho}$ in Step 3 and (2) reduces the cardinality of each E_B in Step 1, where usually the full equivalence class $E_{B,\text{full}} = \{\tilde{B} \in \mathcal{B} : \tilde{B} \sim B\}$ is used, thus leading to a reduction of the dimension of Ξ_E in Step 3 and hence also simplifying Step 4. In fact, one can prove (Michalowsky et al., 2017b, Lemma 4) that for any bracket $B = \tilde{R}_{\psi}(\mathfrak{p}_{i_1,i_r})$ obtained from (2.102), it is $\tilde{B}(x) \equiv 0$ or $\tilde{B}(x) = B(x)$ for any bracket $\tilde{B} \sim B$, i.e., simply put, all equivalent brackets evaluate to zero or to the same vector field. The core property that leads to this result is the fact discussed in Remark 2; we do not elaborate on that here in more detail.

The computation procedure presented beforehand still includes several degrees of freedom, namely the specific choice of frequencies in Step 2 as well as the scalings $\beta_E, \beta_{E,\omega}$ in Step 4. While the conditions on the frequencies are not hard to satisfy and in fact, are not restrictive, it turns out that their choice is crucial in practical implementations. There is still no constructive way of choosing "good" frequencies that we are aware of in the literature. The situation is similar as it comes to the choice of scalings, but here a heuristic way of how to choose them is to distribute the energy of the approximating inputs among different admissible input vector fields ϕ_k. In this spirit, we suggest decreasing the amplitudes of the approximating inputs entering in the primal variables, which will lead to an increase of the amplitudes of the inputs entering in the dual variables. Our simulations results indicate that this procedure usually leads to a better transient and asymptotic behavior of the primal variables, which we are typically most interested in.

C.1.2 Input Strict Passivity of $\mathcal{H}_2$ in Figure 2.12

We first note that, by the Positive Real Lemma (see, e.g., Khalil (2002, Lemma 6.2)), given the pair (A_v, B_v) is controllable and the pair (A_v, C_v) is observable, then the linear system $(A_v, B_v, C_v, 0)$ is passive if and only if there exists a matrix $P = P^\top \succ 0$ of suitable dimension such that

$$PA_v + A_v^\top P \preceq 0 \qquad PB_v = C_v^\top. \tag{C.3}$$

Since the equality constraints are affine according to Assumption 1, there exist $A_{\text{eq}} \in \mathbb{R}^{n_{\text{eq}} \times n_{\text{eq}}}, b_{\text{eq}} \in \mathbb{R}^{n_{\text{eq}}}$ such that $a(z) = Az + b_{\text{eq}}$. Utilizing this specific structure of the equality constraint as well as the Karush-Kuhn-Tucker conditions fulfilled by the saddle point $(z^\star, \nu^\star)$

$$a(z^\star) = A_{\text{eq}} z^\star - b_{\text{eq}} = 0, \qquad \nabla_x L(z^\star, \nu^\star) = \nabla H(z^\star) + A_{\text{eq}}^\top \nu^\star = 0, \tag{C.4}$$

the dynamics $\mathcal{H}_2$ can equivalently be written as

$$\dot{\mu} = A_v \mu + B_v A_{\text{eq}} u_2 \tag{C.5a}$$

$$y_2 = \nabla H(u_2 + z^\star) + A_{\text{eq}}^\top C_v \mu - \nabla H(z^\star). \tag{C.5b}$$

Consider the storage function candidate $V_2(\mu) = \mu^\top P \mu$. Then, utilizing (C.3), the derivative along the solutions of (C.5) fulfills

$$\begin{aligned} \dot{V}_2(\mu) &= \mu^\top (PA_v + A_v^\top P)\mu + \mu^\top C_v^\top A_{\text{eq}} u_2 \\ &\leq u_2^\top \left(y_2 - \nabla H(u_2 + z^\star) + H(z^\star) \right) \end{aligned} \tag{C.6}$$

By strong convexity of H we then infer by (A.24) that

$$\dot{V}_2(\mu) \leq u_2^\top y_2 + m \|u_2\|^2, \tag{C.7}$$

hence input strict passivity of $\mathcal{H}_2$ follows.

C.2 Additional Material to Chapter 3

C.2.1 Comparison to Freeman (2018)

In the following we compare the set of multipliers derived in Freeman (2018) to the multipliers defined in (3.47). For the sake of simplicity we limit ourselves to the case of monotone uncertainties Δ; we emphasize that the same applies to general slope-restricted uncertainties with minor adaptations. Let $M = \text{Toep}([m_j]_{j \in \{-\ell_-, \ldots, \ell_+\}})$, $\bar{M} = \text{Toep}([\bar{m}_j]_{j \in \{-\ell_-, \ldots, \ell_+\}})$, $\ell_+, \ell_- \in \mathbb{N}$, be two Toeplitz operators $M, \bar{M} : \ell_f^{\,1} \to \ell_e^1$. In contrast to Freeman (2018), we consider only finite Toeplitz operators here; the following arguments also apply to infinite operators. In Freeman (2018), the author considers operators $\bar{M}$ with the property

$\bar{M}y = y - h * y$ for all $y \in \ell_f^1$, where $*$ is the convolution operator and $h(z) = \sum_{j=-\ell_-}^{\ell_+} h_j z^{-j}$, i.e., $\bar{M}$ defines the transfer function $E_{\bar{M}}(z) = 1 - h(z)$. It is then shown that if

$$h_j \geq 0 \text{ for } j \in \{-\ell_-, \ldots, \ell_+\} \tag{C.8a}$$

$$\sum_{j=-\ell_-}^{\ell_+} h_j \max\{1, \rho^{-2j}\} \leq 1, \tag{C.8b}$$

then

$$\langle \bar{M}y, \Delta(y)\rangle_w = \langle \rho_- \bar{M}y, \rho_- \Delta(y)\rangle \geq 0 \tag{C.9}$$

for all $y \in \ell_{2,\rho}$, where $\langle x, y\rangle_w = \langle \rho_- x, \rho_- y\rangle$ denotes the weighted inner product on $\ell_{2,\rho}$. Condition (C.8) is equivalently formulated in the parameters $\bar{m}_j$, $j \in \{-\ell_-, \ldots, \ell_+\}$, as

$$\bar{m}_j \leq 0 \text{ for } j \in \{-\ell_-, \ldots, \ell_+\} \setminus \{0\} \text{ and } \bar{m}_0 \geq -1 \tag{C.10a}$$

$$\sum_{j=-\ell_-}^{\ell_+} \bar{m}_{-j} \max\{1, \rho^{-2j}\} \geq 0. \tag{C.10b}$$

Similarly, in Lemma 8 we have shown that if

$$m_j \leq 0 \text{ for } j \in \{-\ell_-, \ldots, \ell_+\} \setminus \{0\} \tag{C.11a}$$

$$\sum_{j=-\ell_-}^{\ell_+} \rho^{-j} m_j \geq 0 \tag{C.11b}$$

$$\sum_{j=-\ell_-}^{\ell_+} \rho^{j} m_j \geq 0. \tag{C.11c}$$

then

$$\langle My, \Delta_\rho(y)\rangle = \langle My, \rho_- \Delta(\rho_+ y)\rangle = \langle M\rho_- \rho_+ y, \rho_- \Delta(\rho_+ y)\rangle \geq 0 \tag{C.12}$$

for all $y \in \ell_2$, i.e., equivalently,

$$\langle M\rho_- y, \rho_- \Delta(y)\rangle \geq 0 \tag{C.13}$$

for all $y \in \ell_{2,\rho}$. We hence note that the operators $\bar{M}$ and M in (C.9) and (C.13) are related by $\rho_- \bar{M} = M\rho_-$ and, therefore,

$$\bar{m}_j = \rho^{-j} m_j \qquad \text{for } j \in \{-\ell_-, \ldots, \ell_+\}. \tag{C.14}$$

We are now ready to compare the conditions (C.10) from Freeman (2018) with the proposed conditions (C.11). To this end, we express (C.10) in terms of m_j instead of $\bar{m}_j$, i.e., we have

$$\rho^{-j} m_j \leq 0 \text{ for } j \in \{-\ell_-, \ldots, \ell_+\} \setminus \{0\} \text{ and } m_0 \geq -1 \tag{C.15a}$$

$$\sum_{j=-\ell_-}^{\ell_+} \rho^{j} m_{-j} \max\{1, \rho^{-2j}\} \geq 0. \tag{C.15b}$$

Note that with $\rho \in (0,1)$ this holds if and only if

$$m_j \le 0 \text{ for } j \in \{-\ell_-, \dots, \ell_+\} \setminus \{0\} \tag{C.16a}$$

$$\sum_{j=-\ell_-}^{\ell_+} \rho^j m_j + \sum_{j=-\ell_-}^{\ell_+} \rho^{-j} m_j = \sum_{j=-\ell_-}^{\ell_+} \rho^{-|j|} m_j \ge 0. \tag{C.16b}$$

We first note that (C.16a) is the same as (C.11a) and next show that (C.16b) implies (C.11b), (C.11c). To this end, observe that for any $j \in \mathbb{Z} \setminus \{0\}$ we have

$$\rho^{|j|} m_j \ge m_j \ge \rho^{-|j|} m_j \tag{C.17}$$

since $\rho \in (0,1)$ and $m_j \le 0$. Consequently,

$$\sum_{j=1}^{\ell_+} \rho^{-j} m_j \le \sum_{j=1}^{\ell_+} \rho^j m_j \tag{C.18a}$$

$$\sum_{j=-\ell_-}^{0} \rho^j m_j = \sum_{j=-\ell_-}^{0} \rho^{-|j|} m_j \le \sum_{j=-\ell_-}^{0} \rho^{|j|} m_j = \sum_{j=-\ell_-}^{0} \rho^{-j} m_j. \tag{C.18b}$$

Thus, (C.16b) implies (C.11b) by (C.18b) and (C.11c) by (C.18a). The converse is in general not true as we will show next. For the sake of a clearer presentation, we suppose that $\ell_- = \ell_+$. Summing up the two inequalities (C.11b), (C.11c), we obtain

$$m_0 + \tfrac{1}{2} \sum_{i=1}^{\ell_-} (\rho^i + \rho^{-i})(m_i + m_{-i}) \ge 0. \tag{C.19}$$

Note further that (C.16b) is equivalently formulated as

$$m_0 + \sum_{i=1}^{\ell_-} \rho^{-i}(m_i + m_{-i}) \ge 0. \tag{C.20}$$

Suppose now that

$$m_0 = -\tfrac{1}{2} \sum_{i=1}^{\ell_-} (\rho^i + \rho^{-i})(m_i + m_{-i}) \tag{C.21}$$

which clearly fulfills (C.19). However, it is

$$-\tfrac{1}{2} \sum_{i=1}^{\ell_-} (\rho^i + \rho^{-i})(m_i + m_{-i}) + \sum_{i=1}^{\ell_-} \rho^{-i}(m_i + m_{-i}) = \sum_{i=1}^{\ell_-} (\rho^{-i} - \rho^i)(m_i + m_{-i}) \le 0 \tag{C.22}$$

and hence m_0 does fulfill (C.20) only if $m_i = m_{-i} = 0$ for $i = 1, 2, \dots, \ell_-$. We hence conclude that the conditions on the anticausal parts proposed in the present thesis are less restrictive than those derived in Freeman (2018). We illustrate the potential benefits by means of

	ℓ_-	ℓ_+	ρ
Purely causal multipliers	1	0	0.977
Anticausal multipliers ((Freeman, 2018))	1	3	0.977
Anticausal multipliers (Theorem 6)	1	3	0.871

Table C.1. Comparison of the resulting convergence rate bounds employing different approaches.

an example. Since anticausal multipliers do not yield an improvement of convergence rate guarantees for the optimization algorithm analysis problem, we consider a numerical example unrelated to the problem at hand. More precisely, we consider a stable linear system described by the following transfer function

$$G(z) = \frac{z + 0.8111}{z^4 + 1.552z^3 + 0.6995z^2 + 0.06042z - 0.01241} \tag{C.23}$$

in feedback with a static, slope-restricted uncertainty Δ in the sector $(0,1)$, i.e., $\Delta \in \mathbf{\Delta}(0,1)$. Employing Theorem 7, we then determine upper bounds on the exponential convergence rates. The resulting convergence rates are displayed in Table C.1. Compared to the multipliers introduced in Freeman (2018), the multipliers introduced in the present thesis lead to an improvement of 10.85%.

C.2.2 State-Space Realizations

In the following we provide explicit state-space realizations of all transfer functions as required for implementation; an overview is given in Table C.2.

State-space realizations of ψ_-, ψ_+. The state-space realization of ψ_+ as defined in (3.50) is given by

$$A_+ = \begin{bmatrix} 0 & 1 & 0 & \cdots \\ \vdots & \ddots & \ddots & \\ \vdots & & 0 & 1 \\ 0 & \cdots & \cdots & 0 \end{bmatrix}, \quad B_+ = \begin{bmatrix} 0 \\ \vdots \\ 0 \\ 1 \end{bmatrix}, \quad C_+ = \begin{bmatrix} 0 & \cdots & 1 \\ & \iddots & \\ & 1 & \\ 1 & 0 \cdots & 0 \end{bmatrix}, \quad D_+ = \begin{bmatrix} 0 \\ \vdots \\ 0 \\ 0 \end{bmatrix}. \tag{C.24a}$$

For ψ_- as defined in (3.50), A_-, B_- have exactly the same structure as A_+, B_+ but may be of different size, while $C_- = I_{\ell_-}$ and $D_- = 0$.

State-space realization of ψ_Δ. The state-space realization of ψ_Δ as defined in (3.53) is given by

$$A_\Delta = \begin{bmatrix} A_- & 0 \\ 0 & A_+ \end{bmatrix} \otimes I_p, \qquad B_\Delta = \begin{bmatrix} B_- \otimes I_p & 0 \\ 0 & B_+ \otimes I_p \end{bmatrix} \widehat{W}, \tag{C.25a}$$

$$C_\Delta = \left[\begin{array}{cc} 0 & 0 \\ 0 & 0 \\ \hline C_- & 0 \\ 0 & 0 \\ \hline 0 & 0 \\ 0 & C_+ \end{array}\right] \otimes I_p, \quad D_\Delta = \left[\begin{array}{cc} I_p & 0 \\ 0 & I_p \\ \hline D_- & 0 \\ 0 & I_p \\ \hline I_p & 0 \\ 0 & D_+ \end{array}\right]. \tag{C.25b}$$

State-space realization of ψ_c. The state-space realization of ψ_c as defined in (3.54) is given by

$$A_c(\rho) = \begin{bmatrix} A_\Delta & \rho^{-1} B_\Delta \begin{bmatrix} C \\ 0 \end{bmatrix} \\ 0 & \rho^{-1} A_{\text{nom}} \end{bmatrix}, \quad B_c = \begin{bmatrix} B_\Delta \begin{bmatrix} 0 \\ I \end{bmatrix} \\ B \end{bmatrix}, \tag{C.26a}$$

$$C_c(\rho) = \begin{bmatrix} C_\Delta & \rho^{-1} D_\Delta \begin{bmatrix} C \\ 0 \end{bmatrix} \end{bmatrix}, \quad D_c = D_\Delta \begin{bmatrix} 0 \\ I \end{bmatrix} \tag{C.26b}$$

with $A_{\text{nom}} = A + mBC$.

State-space realization of ψ_c. We first note that

$$\left[\begin{array}{c|c} G_{yw} & G_{yw_p} \\ I_p & 0 \\ \hline 0 & I_{n_{w_p}} \\ G_{y_p w} & G_{y_p w_p} \end{array}\right] \sim \left(\begin{array}{c|c} A_1 & B_1 \\ \hline C_1 & D_1 \end{array}\right) = \left(\begin{array}{c|c} A_{\text{nom}} & \left[\begin{array}{c|c} B & B_p \end{array}\right] \\ \hline \left[\begin{array}{c} C \\ 0 \\ \hline 0 \\ C_p \end{array}\right] & \left[\begin{array}{c|c} 0 & 0 \\ I_p & 0 \\ \hline 0 & I_{n_{w_p}} \\ 0 & 0 \end{array}\right] \end{array}\right) \tag{C.27}$$

with $A_{\text{nom}} = A + mBC$. Similarly, utilizing that $\psi_p = D_{\psi_p}$ is constant, we have

$$\left[\begin{array}{c|c} \psi_\Delta & 0 \\ \hline 0 & \psi_p \end{array}\right] \sim \left(\begin{array}{c|c} A_2 & B_2 \\ \hline C_2 & D_2 \end{array}\right) = \left(\begin{array}{c|c} A_\Delta & \left[\begin{array}{c|c} B_\Delta & 0 \end{array}\right] \\ \hline \left[\begin{array}{c} C_\Delta \\ \hline 0 \end{array}\right] & \left[\begin{array}{c|c} D_\Delta & 0 \\ \hline 0 & D_{\psi_p} \end{array}\right] \end{array}\right) \tag{C.28}$$

Transfer function	Definition	State-space realization
$\psi_+ \in \mathcal{RH}_\infty^{\ell_+\times 1}$	(3.50)	(A_+, B_+, C_+, D_+)
$\psi_- \in \mathcal{RH}_\infty^{\ell_-\times 1}$	(3.50)	(A_-, B_-, C_-, D_-)
$\psi_\Delta \in \mathcal{RH}_\infty^{p(\ell_-+\ell_++4)\times 2p}$	(3.53)	$(A_\Delta, B_\Delta, C_\Delta, D_\Delta)$
$\psi_c \in \mathcal{RH}_\infty^{p(\ell_-+\ell_++4)\times p}$	(3.54)	(A_c, B_c, C_c, D_c)
$\boldsymbol{\psi}_c \in \mathcal{RH}_\infty^{(p(4+\ell_-+\ell_+)+q)\times 2(p+n_{w\mathrm{p}})}$	(3.61)	$(\boldsymbol{A}_c, \boldsymbol{B}_c, \boldsymbol{C}_c, \boldsymbol{D}_c)$

Table C.2. Overview of all required state-space realizations.

Note that $D_{\psi_\mathrm{p}} = \begin{bmatrix} 0_{n_{y\mathrm{p}}\times n_{w\mathrm{p}}} & I_{n_{y\mathrm{p}}} \end{bmatrix}$ in the case of H_2-performance. By standard rules for series connections, the state-space realization of $\boldsymbol{\psi}_c$ as defined in (3.61) is then given by

$$\boldsymbol{A}_c = \left[\begin{array}{c|c} A_2 & B_2C_1 \\ \hline 0 & A_1 \end{array}\right] = A_c(1), \qquad \boldsymbol{B}_c = \left[\begin{array}{c} B_2D_1 \\ \hline B_1 \end{array}\right] = \left[\begin{array}{c|c} B_c & \begin{bmatrix} 0 \\ B_\mathrm{p} \end{bmatrix} \end{array}\right], \tag{C.29a}$$

$$\boldsymbol{C}_c = \begin{bmatrix} C_2 & D_2C_1 \end{bmatrix} = \left[\begin{array}{c} C_c(1) \\ \hline \begin{bmatrix} 0 & D_{\psi_\mathrm{p}} \begin{bmatrix} 0 \\ C_\mathrm{p} \end{bmatrix} \end{bmatrix} \end{array}\right], \qquad \boldsymbol{D}_c = D_2D_1 = \left[\begin{array}{c|c} D_c & 0 \\ \hline 0 & D_{\psi_\mathrm{p}} \begin{bmatrix} I_{n_{w\mathrm{p}}} \\ 0 \end{bmatrix} \end{array}\right]. \tag{C.29b}$$

C.2.3 Derivation of (3.72)

Consider (3.55a) and note that we may rewrite its left-hand side as

$$U(A_c(\rho), B_c, C_c(\rho), D_c, P, M_\Delta) - [\star]^\top NN^\top PNN^\top \begin{bmatrix} A_c(\rho) & B_c \end{bmatrix} \prec 0. \tag{C.30}$$

Utilizing that $P_{22} = N^\top PN \succ 0$ and employing Schur complements, (3.72) immediately follows.

List of Symbols

The following list presents an overview of the most frequently used symbols and acronyms. Precise definitions are given at the symbol's first appearance or in Appendix A.

General Notation and Symbols

Sets

$\mathbb{N}$	Set of non-negative integers.
$\mathbb{N}_{>0}$	Set of positive integers.
$\mathbb{Z}$	Set of integers.
$\mathbb{R}$	Set of reals.
$\mathbb{R}_{\geq 0}$	Set of non-negative reals.
$\mathbb{R}_{>0}$	Set of positive reals.
$\mathbb{C}$	Set of complex numbers.
$\mathbb{T}$	Unit circle in the complex plane, i.e., $\mathbb{T} = \{z \in \mathbb{C} : \|z\| = 1\}$.

Matrices and Vectors

I_n	The $n \times n$ identity matrix. We often do not specify the dimension.
$0_{n\times m}$	The $n \times m$ matrix of zeros. We often do not specify the dimension.
$[a_{ij}]_{i=1,\ldots,n,\, j=1,\ldots,m}$	A $n \times m$ matrix with the (i,j)th entry being $a_{ij} \in \mathbb{R}$.
$[x_i]_{i\in S}$	The stacked column vector of all x_i, $i \in S \subset \mathbb{N}_{>0}$, ordered by the index i.
$\operatorname{diag}(m_1,\ldots,m_k)$	Diagonal matrix with entries $m_1,\ldots,m_k \in \mathbb{R}$ on the diagonal.
$\operatorname{blkdiag}(M_1,\ldots,M_k)$	Block diagonal matrix with blocks $M_1,\ldots,M_k$ on the diagonal.
e_i	The ith unit vector, i.e., a vector with the ith entry being equal to one and all others being equal to zero.
$\mathbf{1}$	A vector with all entries being equal to one.
$A > 0$ $(A \geq 0)$	All elements of $A \in \mathbb{R}^{n\times p}$ are positive (non-negative).
$A < 0$ $(A \leq 0)$	All elements of $A \in \mathbb{R}^{n\times p}$ are negative (non-positive).
$A \succ 0$ $(A \succeq 0)$	$A \in \mathbb{R}^{n\times n}$ is positive (semi-)definite.
$A \prec 0$ $(A \preceq 0)$	$A \in \mathbb{R}^{n\times n}$ is negative (semi-)definite.
$\otimes$	Kronecker matrix product.

Function Spaces

$\mathcal{C}^p$	The set of p-times continuously differentiable functions.
$\mathcal{S}_{m,L}$	The set of twice continuously differentiable functions being m-strongly convex and having L-Lipschitz continuous gradient.

Notation and Symbols Specific for Chapter 2

Graphs and Paths

$\mathcal{G}$	A directed graph.
$\mathcal{V}$	Set of nodes.
$\mathcal{E}$	Set of edges.
G	Laplacian of a graph.
$\mathbf{A}$	Adjacency matrix of a graph.
D	Out-degree matrix of a graph.
$\mathcal{N}_{\mathcal{G}}(i)$	Set of out-neighboring nodes of node i in the graph $\mathcal{G}$.
$\mathfrak{p}_{i,j} = \langle i \mid \ldots \mid j \rangle$	A path from node i to node j.
$\ell(\mathfrak{p}_{i,j})$	Length of the path $\mathfrak{p}_{i,j}$.

Saddle-Point Dynamics and Distributed Implementation

L	Lagrangian corresponding to an optimization problem.
z	Primal variable.
ν	Dual variable corresponding to the equality constraints.
λ	Dual variable corresponding to the inequality constraints.
x	Complete state vector.
$\mathbf{x}_i$	State of the ith agent.
$\mathcal{I}_{\mathrm{eq}}(i)$	Set of equality constraints associated to agent i.
$\mathcal{I}_{\mathrm{ineq}}(i)$	Set of inequality constraints associated to agent i.
$\mathcal{I}(i)$	Set of indices of the complete state vector associated to the ith agent, where $\mathcal{I}$ is called an agent-to-state assignment.
$\Phi_{\mathrm{all}}^{\mathrm{adm}}(\mathcal{G},\mathcal{I})$	Set of all vector fields admissible for a graph $\mathcal{G}$ with respect to the agent-to-state assignment $\mathcal{I}$.
$\Phi^{\mathrm{adm}}(\mathcal{G},\mathcal{I})$	Subset of admissible vector fields in $\Phi_{\mathrm{all}}^{\mathrm{adm}}(\mathcal{G},\mathcal{I})$ that are required for the admissible Lie bracket representations.
$f^{\mathrm{adm}}, f^{\neg\mathrm{adm}}$	Admissible and non-admissible part of a vector field f.

Lie Brackets

$[\phi_1, \phi_2]$	Lie bracket of two vector fields ϕ_1, ϕ_2.
$\delta(B)$	Degree of a Lie bracket B.
$\mathrm{left}(B)$	Left factor of a Lie bracket B.
$\mathrm{right}(B)$	Right factor of a Lie bracket B.
$\mathcal{PH}(\Phi)$	P. Hall basis of a set of vector fields Φ.
$\mathcal{LB}r(\Phi)$	Set of Lie brackets built from the set of vector fields Φ.
$\mathcal{FB}r(\Phi)$	Set of formal brackets built from the set of indeterminants Φ.

Notation and Symbols Specific for Chapter 3

Transfer Functions

$\mathcal{RL}_\infty^{n\times m}$ (resp. $\mathcal{RH}_\infty^{n\times m}$)	The set of all real-rational transfer matrices of dimension $n \times m$ having no poles on the unit circle (resp., having all poles in the open unit disk).
$G \overset{\mathbb{S}}{\succ} 0$	$G \in \mathcal{RL}_\infty^{n\times n}$ fulfills $G(z) > 0$ for all $z \in \mathbb{S} \subset \mathbb{C}$.

Sequences and Operators

$(q_k)_{k\in\mathbb{N}} = (q_0, q_1, \dots)$	A sequence $q : k \mapsto q_k$ with $q_k \in \mathbb{R}^p$, $p \in \mathbb{N}_{>0}$.
ℓ_e^p	Set of all one-sided sequences on $\mathbb{R}^p$, i.e, $\ell_e^p = \{(q_k)_{k\in\mathbb{N}} \mid q_k \in \mathbb{R}^p\}$.
ℓ_e	Collection of all ℓ_e^p, $p = 1, 2, \dots$.
ℓ_f^p (resp. ℓ_f)	Set of all finitely supported sequences in ℓ_e^p (resp. ℓ_e).
ℓ_2^p (resp. ℓ_2)	Set of all square-summable sequences in ℓ_e^p (resp. ℓ_e).
$\langle u, v\rangle$	Standard inner product on ℓ_2, i.e., $\langle u, v\rangle = \sum_{i=0}^{\infty} u_i^\top v_i$, $u, v \in \ell_2^p$.
$\lVert u\rVert_{\ell_2}$	Induced norm on ℓ_2, i.e., $\lVert u\rVert = \sqrt{\langle u, u\rangle}$, $u \in \ell_2^p$.
id	Identity operator mapping a sequence in ℓ_e^p to itself.
ρ_+, ρ_-	Time-varying signal transformations (exponential weightings).
$\ell_{2,\rho}^p$ (resp. $\ell_{2,\rho}$)	Set of all sequences in ℓ_2^p (resp. ℓ_2) that decay exponentially with rate $\rho \in (0, 1)$.
O_T	Truncation operator truncating a sequence at step T, i.e., for $q \in \ell_e^p$ we have $O_T(q) = (q_0, q_1, \dots, q_T, 0, 0, \dots)$.
$\widehat{q}$	The one-sided z-transform of a sequence $q \in \ell_2$.
ϕ_G	Linear Toeplitz operator $\phi_G : \ell_2^m \to \ell_2^n$ corresponding to a transfer matrix $\mathcal{G} \in \mathcal{RL}_\infty^{n\times m}$.

State-Space Representations

$G \sim (A, B, C, D)$ or $G \sim \left(\begin{array}{c\|c} A & B \\ \hline C & D \end{array}\right)$	The transfer matrix $G(z) = C(zI - A)^{-1}B + D$ admits the state-space representation (A, B, C, D).

Uncertainties and Integral Quadratic Constraints

Δ	Uncertain operator.
$\boldsymbol{\Delta}$	Set of uncertain operators.
Π	IQC multiplier in the frequency domain, $\Pi \in \mathcal{RL}_\infty$.
$\boldsymbol{\Pi}$	Set of IQC multipliers.

Acronyms

IQC	Integral quadratic constraint
LMI	Linear matrix inequality
BMI	Bilinear matrix inequality
FDI	Frequency domain inequality
MPC	Model predictive control
HB	Heavy Ball Method
GD	Gradient Descent Algorithm
NM	Nesterov's Method
TMM	Triple Momentum Method

Bibliography

Antipin, A. S. (1994). Minimization of convex functions on convex sets by means of differential equations. *Differential equations, 30*(9), 1365–1375.
Ariyur, K. B., & Krstić, M. (2003). *Real-time optimization by extremum-seeking control.* John Wiley & Sons.
Åström, K., & Wittenmark, B. (1995). *Adaptive Control.* Dover Publications.
Aybat, N. S., Fallah, A., Gürbüzbalaban, M., & Ozdaglar, A. (2019). Robust accelerated gradient methods for smooth strongly convex functions. *ArXiv e-prints arXiv:1805.10579.* (https://arxiv.org/abs/1805.10579)
Belabbas, M.-A. (2013). Sparse stable systems. *Systems & Control Letters, 62*(10), 981–987.
Ben Tal, A., Tsibulevskii, M., & Yusefovich, I. (1992). Modified barrier methods for constrained and minimax problems. *Technical Report, Optim. Lab., Technion.*
Bhaya, A., & Kaszkurewicz, E. (2006). *Control perspectives on numerical algorithms and matrix problems* (Vol. 10). SIAM.
Biggs, N. (1993). *Algebraic graph theory.* Cambridge University Press.
Boczar, R., Lessard, L., Packard, A., & Recht, B. (2017). Exponential stability analysis via Integral Quadratic Constraints. *ArXiv e-prints arXiv:1706.01337.* (https://arxiv.org/abs/1706.01337)
Boczar, R., Lessard, L., & Recht, B. (2015). Exponential convergence bounds using integral quadratic constraints. In *Proc. 54th IEEE Conference on Decision and Control (CDC)* (pp. 7516–7521).
Bourbaki, N. (1998). *Lie groups and Lie algebras: Chapters 1–3.* Springer.
Boyd, S., El Ghaoui, L., Feron, E., & Balakrishnan, V. (1994). *Linear matrix inequalities in system and control theory* (Vol. 15). SIAM.
Boyd, S., Parikh, N., Chu, E., Peleato, B., & Eckstein, J. (2011). Distributed optimization and statistical learning via the alternating direction method of multipliers. *Foundations and Trends in Machine Learning, 3*(1), 1–122.
Boyd, S., & Vandenberghe, L. (2004). *Convex Optimization.* Cambridge University Press.
Bregman, L. (1967). The relaxation method of finding the common point of convex sets and its application to the solution of problems in convex programming. *USSR computational mathematics and mathematical physics, 7*(3), 200–217.
Brockett, R. W. (1988). Dynamical systems that sort lists, diagonalize matrices and solve linear programming problems. In *Proc. 27th IEEE Conf. Decision and Control* (pp. 799–803).
Brockett, R. W. (1991). Dynamical systems that sort lists, diagonalize matrices, and solve linear programming problems. *"Linear Algebra and its Applications", 146,* 79–91.
Brockett, R. W. (2014). The early days of geometric nonlinear control. *Automatica, 50*(9), 2203–2224.
Bullo, F., Cortés, J., & Martínez, S. (2009). *Distributed control of robotic networks.* Princeton University Press.
Chen, X., Belabbas, M.-A., & Başar, T. (2015). Controllability of formations over directed graphs. In *Proc. 54th IEEE Conf. Decision and Control (CDC)* (pp. 4764–4769).
Costello, Z., & Egerstedt, M. (2014). The degree of nonholonomy in distributed computations. In *Proc. 53rd IEEE Conf. Decision and Control (CDC)* (pp. 6092–6098).
Cyrus, S., Hu, B., Van Scoy, B., & Lessard, L. (2018). A robust accelerated optimization algorithm for

strongly convex functions. In *Proc. American Control Conf. (ACC)* (pp. 1376–1381).
D'Amato, F. J., Rotea, M. A., Megretski, A., & Jönsson, U. T. (2001). New results for analysis of systems with repeated nonlinearities. *Automatica, 37*(5), 739–747.
Desoer, C. A., & Vidyasagar, M. (1975). *Feedback systems: Input-output properties* (Vol. 55). SIAM.
Draper, C. S., & Li, Y. T. (1951). *Principles of optimalizing control systems and an application to the internal combustion engine.* American Society of Mechanical Engineers.
Drori, Y., & Taylor, A. B. (2018). Efficient first-order methods for convex minimization: A constructive approach. *ArXiv e-prints arXiv:1803.05676.* (https://arxiv.org/abs/1803.05676)
Drori, Y., & Teboulle, M. (2014). Performance of first-order methods for smooth convex minimization: A novel approach. *Mathematical Programming, 145*(1), 451–482.
Dürr, H.-B. (2015). *Constrained Extremum Seeking: A Lie bracket and singular perturbation approach* (PhD Thesis). University of Stuttgart.
Dürr, H.-B., & Ebenbauer, C. (2012). On a class of smooth optimization algorithms with applications in control. In *Proc. 4th IFAC Conf. Nonlinear Model Predictive Control (NMPC)* (pp. 291–298).
Dürr, H.-B., Stanković, M. S., Ebenbauer, C., & Johansson, K. H. (2013). Lie bracket approximation of extremum seeking systems. *Automatica, 49*(6), 1538–1552.
Dürr, H.-B., Zeng, C., & Ebenbauer, C. (2013). Saddle point seeking for convex optimization problems. In *Proc. 9th IFAC Symp. Nonlinear Control Systems (NOLCOS)* (pp. 540–545).
Ebenbauer, C., Michalowsky, S., Grushkovskaya, V., & Gharesifard, B. (2017). Distributed optimization over directed graphs with the help of Lie brackets. In *Proc. 20th IFAC World Congress* (pp. 15908–15913).
Fazlyab, M., Morari, M., & Preciado, V. M. (2018). Design of first-order optimization algorithms via sum-of-squares programming. In *Proc. 57th IEEE Conf. Decision and Control (CDC)* (pp. 4445–4452).
Fazlyab, M., Ribeiro, A., Morari, M., & Preciado, V. (2018). Analysis of optimization algorithms via Integral Quadratic Constraints: Nonstrongly convex problems. *SIAM Journal on Optimization, 28*(3), 2654–2689.
Feijer, D., & Paganini, F. (2010). Stability of primal–dual gradient dynamics and applications to network optimization. *Automatica, 46*(12), 1974–1981.
Feiling, J., Zeller, A., & Ebenbauer, C. (2018). Derivative-Free Optimization Algorithms Based on Non-Commutative Maps. *IEEE Control Systems Letters, 2*(4), 743–748.
Feller, C., & Ebenbauer, C. (2017a). Relaxed logarithmic barrier function based model predictive control of linear systems. *IEEE Transactions on Automatic Control, 62*(3), 1223–1238.
Feller, C., & Ebenbauer, C. (2017b). A stabilizing iteration scheme for model predictive control based on relaxed barrier functions. *Automatica, 80*, 328–339.
Fetzer, M., & Scherer, C. W. (2017). Absolute stability analysis of discrete time feedback interconnections. In *Proc. 20th IFAC World Congress* (pp. 8447–8453).
Freeman, R. A. (2018). Noncausal Zames-Falb multipliers for tighter estimates of exponential convergence rates. In *Proc. American Control Conf. (ACC)* (pp. 2984–2989).
Gharesifard, B. (2017). Stabilization of bilinear sparse matrix control systems using periodic inputs. *Automatica, 77*(Supplement C), 239–245.
Gharesifard, B., & Cortés, J. (2014). Distributed continuous-time convex optimization on weight-balanced digraphs. *IEEE Transactions on Automatic Control, 59*(3), 781–786.
Grushkovskaya, V., Zuyev, A., & Ebenbauer, C. (2018). On a class of generating vector fields for the extremum seeking problem: Lie bracket approximation and stability properties. *Automatica, 94*, 151–160.
Guay, M., Vandermeulen, I., Dougherty, S., & Mclellan, P. J. (2015). Distributed extremum-seeking control over networks of dynamic agents. In *Proc. American Control Conf. (ACC)* (pp. 159–164).
Hauser, J., & Saccon, A. (2006). A barrier function method for the optimization of trajectory

functionals with constraints. In *Proc. 45th IEEE Conference on Decision and Control (CDC)* (pp. 864–869).

Hauswirth, A., Bolognani, S., Hug, G., & Dörfler, F. (2019). Timescale separation in autonomous optimization. *arXiv preprint arXiv:1905.06291*. (https://arxiv.org/abs/1905.06291)

Helmke, U., & Moore, J. B. (1994). *Optimization and dynamical systems*. Springer-Verlag.

Hiriart-Urruty, J.-B., & Lemaréchal, C. (2013). *Convex analysis and minimization algorithms I: Fundamentals* (Vol. 305). Springer science & business media.

Hu, B., & Lessard, L. (2017). Dissipativity theory for Nesterov's accelerated method. In *Proc. 34th International Conference on Machine Learning* (pp. 1549–1557).

Hu, B., & Seiler, P. (2016). Exponential decay rate conditions for uncertain linear systems using Integral Quadratic Constraints. *IEEE Transactions on Automatic Control, 61*(11), 3631–3637.

Jury, E., & Lee, B. (1964). On the stability of a certain class of nonlinear sampled-data systems. *IEEE Transactions on Automatic Control, 9*(1), 51–61.

Kao, C. (2012). On stability of discrete-time LTI systems with varying time delays. *IEEE Transactions on Automatic Control, 57*(5), 1243–1248.

Khalil, H. K. (2002). *Nonlinear Systems*. Prentice Hall.

Khatib, O. (1986). Real-time obstacle avoidance for manipulators and mobile robots. In *Autonomous robot vehicles* (pp. 396–404). Springer Science & Business Media.

Koren, Y., & Borenstein, J. (1991). Potential field methods and their inherent limitations for mobile robot navigation. In *Proc. IEEE International Conf. on Robotics and Automation* (pp. 1398–1404).

Kurzweil, J., & Jarník, J. (1987). Limit processes in ordinary differential equations. *Zeitschrift für angewandte Mathematik und Physik, 38*(2), 241–256.

Labar, C., Garone, E., Kinnaert, M., & Ebenbauer, C. (2019). Newton-based extremum seeking: A second-order Lie bracket approximation approach. *Automatica, 105*, 356–367.

Leblanc, M. (1922). Sur l'électrification des chemins de fer au moyen de courants alternatifs de frequence élevée. *Revue Générale de l'Électricité, 12*(8), 275–277.

Lessard, L., Recht, B., & Packard, A. (2016). Analysis and design of optimization algorithms via integral quadratic constraints. *SIAM Journal on Optimization, 26*(1), 57–95.

Lessard, L., & Seiler, P. (2019). Direct synthesis of iterative algorithms with bounds on achievable worst-case convergence rate. *ArXiv e-prints arXiv:1904.09046*. (https://arxiv.org/abs/1904.09046)

Li, C., Qu, Z., & Weitnauer, M. (2015). Distributed extremum seeking and formation control for nonholonomic mobile network. *Systems & Control Letters, 75*, 27–34.

Li, Z., & Canny, J. F. (2012). *Nonholonomic motion planning* (Vol. 192). Springer Science & Business Media.

Liu, W. (1997a). An approximation algorithm for nonholonomic systems. *SIAM J. Control Optim., 35*(4), 1328–1365.

Liu, W. (1997b). Averaging theorems for highly oscillatory differential equations and iterated Lie brackets. *SIAM J. Control Optim., 35*(6), 1989–2020.

Löfberg, J. (2004). YALMIP : A Toolbox for Modeling and Optimization in MATLAB. In *Proc. CACSD Conference*.

Lur'e, A. I., & Postnikov, V. N. (1944). On the theory of stability of control systems. *Applied mathematics and mechanics, 8*(3), 246–248.

Mancera, R., & Safonov, M. G. (2005). All stability multipliers for repeated MIMO nonlinearities. *Systems & Control Letters, 54*(4), 389–397.

Megretski, A., & Rantzer, A. (1997). System analysis via integral quadratic constraints. *IEEE Transactions on Automatic Control, 42*(6), 819–830.

Michalowsky, S., & Ebenbauer, C. (2014). The multidimensional n-th order heavy ball method and its application to extremum seeking. In *Proc. 53rd IEEE Conf. Decision and Control (CDC)* (pp.

2660–2666).

Michalowsky, S., & Ebenbauer, C. (2016). Extremum control of linear systems based on output feedback. In *Proc. 55th IEEE Conf. Decision and Control (CDC)* (pp. 2963–2968).

Michalowsky, S., Gharesifard, B., & Ebenbauer, C. (2017a). Distributed extremum seeking over directed graphs. In *Proc. 56th IEEE Conf. Decision and Control (CDC)* (pp. 2095–2101).

Michalowsky, S., Gharesifard, B., & Ebenbauer, C. (2017b). A Lie bracket approximation approach to distributed optimization over directed graphs. *ArXiv e-prints arXiv:1711.05486.* (https://arxiv.org/abs/1711.05486)

Michalowsky, S., Gharesifard, B., & Ebenbauer, C. (2018). On the Lie bracket approximation approach to distributed optimization: Extensions and limitations. In *Proc. European Control Conf. (ECC)* (pp. 119–124).

Michalowsky, S., Gharesifard, B., & Ebenbauer, C. (2020). A Lie bracket approximation approach to distributed optimization over directed graphs. *Automatica, 112*, 108691.

Michalowsky, S., Scherer, C., & Ebenbauer, C. (2020). Robust and structure exploiting algorithms: An integral quadratic constraint approach. *Int. J. Control*, to appear. (DOI: 10.1080/00207179.2020.1745286)

Mohammadi, H., Razaviyayn, M., & Jovanović, M. R. (2018). Variance amplification of accelerated first-order algorithms for strongly convex quadratic optimization problems. In *Proc. 57th IEEE Conf. Decision and Control (CDC)* (pp. 5753–5758).

Murphy, K. P., & Bach, F. (2012). *Machine Learning: A probabilistic perspective*. MIT Press.

Nash, S. G., Polyak, R., & Sofer, A. (1994). A numerical comparison of barrier and modified barrier methods for large-scale bound-constrained optimization. In *Large scale optimization: State of the art* (pp. 319–338). Kluwer Academic Publishers.

Nedić, A., & Liu, J. (2018). Distributed optimization for control. *Annual Review of Control, Robotics, and Autonomous Systems, 1*(1), 77–103.

Nesterov, Y. (2004). *Introductory lectures on convex optimization* (Vol. 87). Springer Science & Business Media.

Nesterov, Y., & Nemirovskii, A. (1994). *Interior-point polynomial algorithms in convex programming*. SIAM.

Niederländer, S. K., & Cortés, J. (2015). Distributed coordination for separable convex optimization with coupling constraints. In *Proc. 54th IEEE Conf. Decision and Control (CDC)* (pp. 694–699).

The On-Line Encyclopedia of Integer Sequences, A000048. (n.d.). https://oeis.org/A000048. OEIS Foundation Inc. (2017).

The On-Line Encyclopedia of Integer Sequences, A006788. (n.d.). https://oeis.org/A006788. OEIS Foundation Inc. (2017).

O'Shea, R., & Younis, M. (1967). A frequency-time domain stability criterion for sampled-data systems. *IEEE Transactions on Automatic Control, 12*(6), 719-724.

Paganini, F., & Feron, E. (2000). Linear matrix inequality methods for robust H_2 analysis: A survey with comparisons. In *Advances in linear matrix inequality methods in control* (pp. 129–151).

Polyak, B. T. (1987). *Introduction to optimization*. Optimization Software Inc.

Popov, V. M. (1961). On absolute stability of non-linear automatic control systems. *Automatika i Telemekhanika, 22*(8), 961–979.

Reutenauer, C. (2003). Free Lie algebras. *Handbook of Algebra, 3*, 887–903.

Rockafellar, R. T. (1970). *Convex Analysis*. Princeton University Press.

Romer, A., Montenbruck, J. M., & Allgöwer, F. (2017). Sampling strategies for data-driven inference of passivity properties. In *Proc. 56th IEEE Conf. Decision and Control (CDC) Decision and Control (CDC)* (pp. 6389–6394).

Safavi, S., Joshi, B., França, G., & Bento, J. (2018). An explicit convergence rate for Nesterov's method from SDP. In *2018 IEEE International Symposium on Information Theory (ISIT)* (pp. 1560–1564).

Stanley, R. P. (2015). *Catalan Numbers*. Cambridge University Press.

Sternby, J. (1980). Adaptive control of extremum systems. *"Methods and Applications in Adaptive Control: Proc. of an International Symposium"*, 151–160.

Su, W., Boyd, S., & Candes, E. (2014). A differential equation for modeling Nesterov's accelerated gradient method: Theory and insights. In *Advances in Neural Information Processing Systems* (pp. 2510–2518).

Sussmann, H. J., & Liu, W. (1991). Limits of highly oscillatory controls and the approximation of general paths by admissible trajectories. In *Proc. 30th IEEE Conf. Decision and Control (CDC)* (pp. 437–442).

Taylor, A. B., Hendrickx, J. M., & Glineur, F. (2017). Performance estimation toolbox (PESTO): Automated worst-case analysis of first-order optimization methods. In *Proc. 56th IEEE Conf. Decision and Control (CDC)* (pp. 1278–1283).

Touri, B., & Gharesifard, B. (2016). Saddle-point dynamics for distributed convex optimization on general directed graphs. In *Proc. 55th IEEE Conf. Decision and Control (CDC)* (pp. 862–866).

Tsypkin, Y. Z. (1964). A Criterion for Absolute Stability of Automatic Pulse Systems with Monotonic Characteristics of the Nonlinear Element. *Soviet Physics Doklady*, *9*.

Van Scoy, B., Freeman, R. A., & Lynch, K. M. (2018). The fastest known globally convergent first-order method for minimizing strongly convex functions. *IEEE Control Systems Letters*, *2*(1), 49–54.

Veenman, J., Scherer, C. W., & Köroğlu, H. (2016). Robust stability and performance analysis based on integral quadratic constraints. *European J. Control*, *31*, 1–32.

Wang, J., & Elia, N. (2010). Control approach to distributed optimization. In *Proc. 48th Annual Allerton Conf. on Communication, Control, and Computing* (pp. 557–561).

Wang, J., & Elia, N. (2011). A control perspective for centralized and distributed convex optimization. In *Proc. 50th IEEE Conf. Decision and Control (CDC), European Control Conf. (ECC)* (pp. 3800–3805).

Wibisono, A., Wilson, A. C., & Jordan, M. I. (2016). A variational perspective on accelerated methods in optimization. *Proceedings of the National Academy of Sciences*, *113*(47), E7351–E7358.

Willems, J. (1971). *The analysis of feedback systems*. MIT Press.

Willems, J., & Brockett, R. W. (1968). Some new rearrangement inequalities having application in stability analysis. *IEEE Transactions on Automatic Control*, *13*(5), 539–549.

Wilson, A. C., Recht, B., & Jordan, M. I. (2016). A Lyapunov analysis of momentum methods in optimization. *ArXiv e-prints arXiv:1611.02635*. (https://arxiv.org/abs/1611.02635)

Yamashita, S., Hatanaka, T., Yamauchi, J., & Fujita, M. (2018). Passivity-based generalization of primal-dual dynamics for non-strictly convex cost functions. *ArXiv e-prints arXiv:1811.08640*. (https://arxiv.org/abs/1811.08640)

Zames, G., & Falb, P. (1968). Stability conditions for systems with monotone and slope-restricted nonlinearities. *SIAM Journal on Control*, *6*(1), 89–108.

Zhang, J., Seiler, P., & Carrasco, J. (2019). Noncausal FIR Zames-Falb multiplier search for exponential convergence rate. *ArXiv e-prints arXiv:1902.09473*. (https://arxiv.org/abs/1902.09473)

Zhao, J., & Dörfler, F. (2015). Distributed control and optimization in DC microgrids. *Automatica*, *61*, 18–26.